AF594636

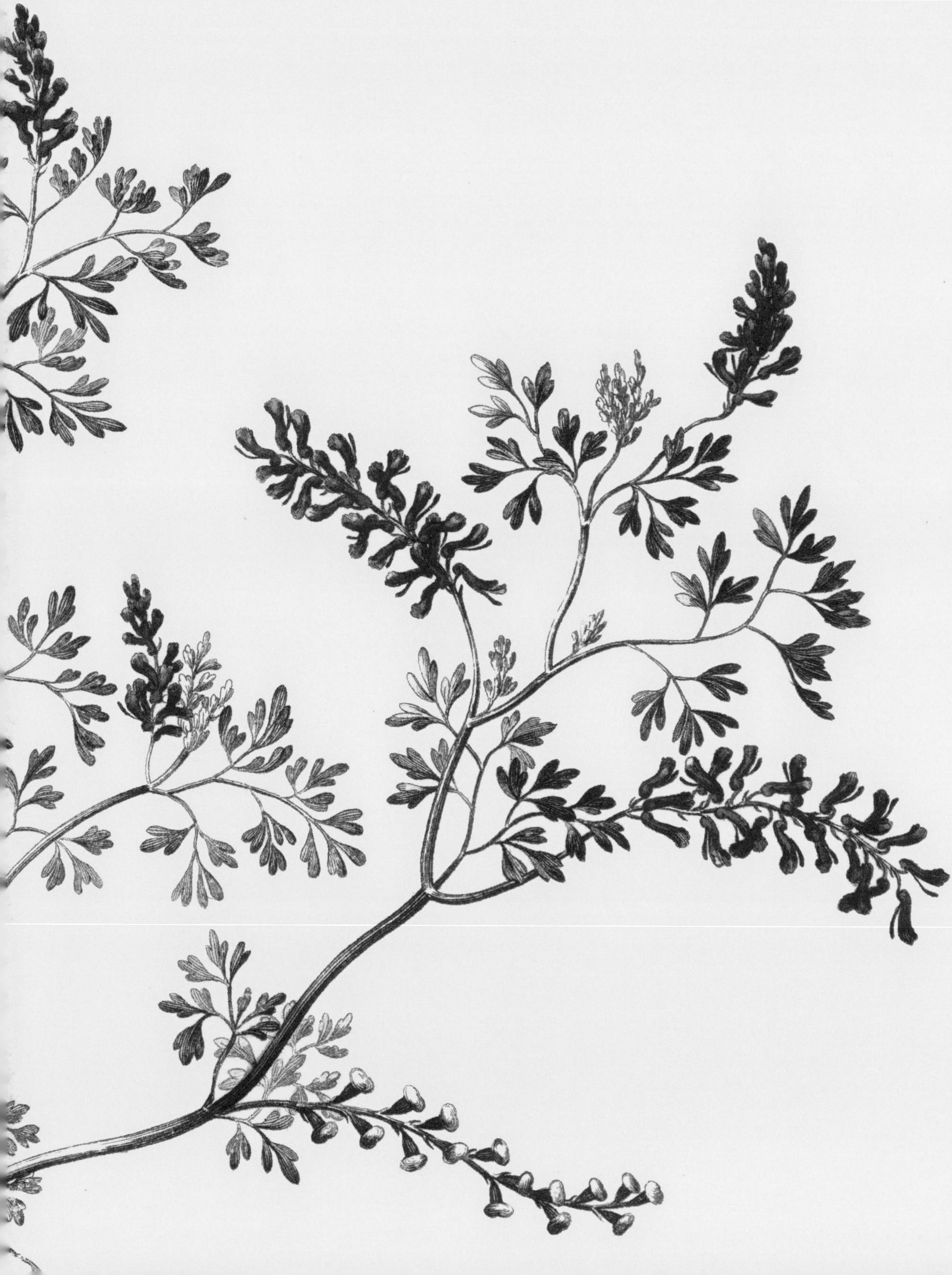

THE *BOTANICAL* CITY

A busy person's guide to the

WONDROUS PLANTS

you can find, eat and grow

IN THE CITY

The Botanical City:
A busy person's guide to the wondrous plants
you can find, eat and grow in the city

First edition, second printing
First published by Hoxton Mini Press in collaboration with
the Royal Botanic Gardens, Kew, London 2020

Reproduced from *Flora Londinensis: or plates and descriptions of such plants as grow wild in the environs of London, etc.* by William Curtis, London, 1775–1798

Written by Hélèna Dove and Harry Adès
Book and cover design by Daniele Roa
Copy-editing by Faith McAllister
Production by Anna De Pascale
Illustration retouching by Becca Jones

ISBN: 978-1-910566-79-4

A CIP catalogue record for this book is available from the British Library.

This book is 100% carbon compensated according to ClimateCalc (climatecalc.eu). Offset purchased from: Stand for Trees.

Printed and bound by Livonia Print, Latvia

www.hoxtonminipress.com

ACKNOWLEDGEMENTS

Kew Publishing would like to thank Fiona Ainsworth, Pei Chu, Paul Little, Anne Marshall, Lynn Parker, Martyn Rix and Kim Walker for their help with this publication. Hoxton Mini Press would like to thank the writers Harry Adès and Hélèna Dove for their incredibly hard work and perseverance, and Gina Fullerlove and Lydia White at Kew Publishing for their faith, support and vision. This has been one of our favourite collaborations to date.

DISCLAIMER

The information in this book is intended to educate, delight and expand the reader's understanding of the diversity of plant life. It does not purport to be, nor is it intended to be, a medical manual or a self-treatment guide, a recipe book or a means of diagnosing poisoning. The authors and publishers make no claims about the efficacy or safety of the plants and compounds mentioned in the book, do not endorse the use of these plants for any of the applications described, do not endorse picking and eating wild plants and mushrooms, and are not responsible for any consequences arising from the use of this information for whatever reason, including curiosity or malicious or illegal intent. The information in this book is not a substitute for advice or medication from a healthcare professional. The authors specifically disclaim all responsibility for any liability, loss, or risk, personal or otherwise which is incurred as a consequence, directly or indirectly, of the use and application of any of the contents of this book.

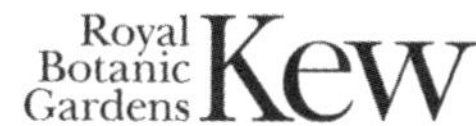

THE *BOTANICAL* CITY

Written by

HÉLÈNA DOVE and *HARRY ADÈS*

With illustrations from
the 18th-century book
FLORA LONDINENSIS
compiled by
WILLIAM CURTIS

HOXTON *MINI* PRESS

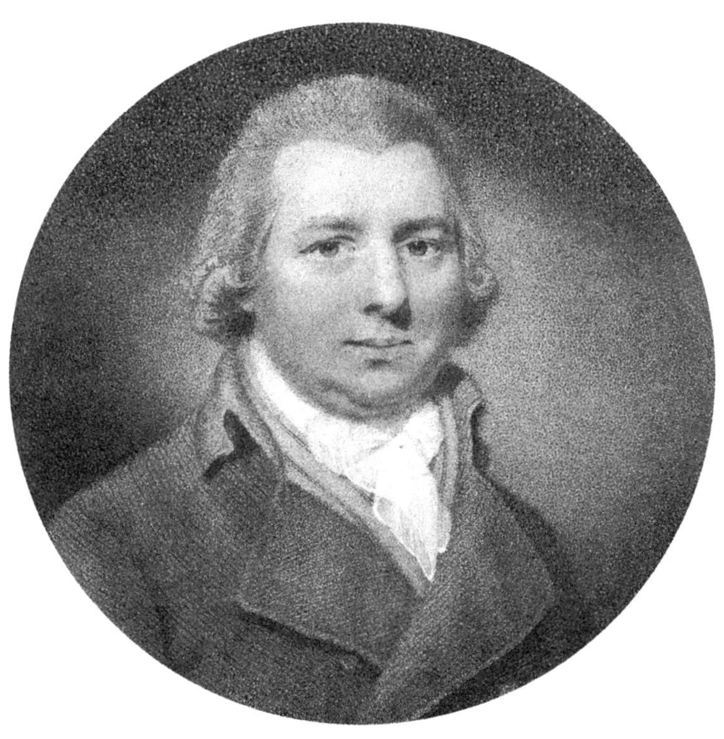

William Curtis (1746–1799) author of
Flora Londinensis and *The Botanical Magazine*

WILLIAM CURTIS AND *FLORA LONDINENSIS*

The book in your hands owes everything to the botanist, William Curtis (1746–1799), and his masterpiece, *Flora Londinensis*. This work, one of the most influential and beautiful botanical studies of its age, took Curtis most of his adult life to complete and very nearly ruined him in the process.

Published serially between 1775 and 1798, it featured detailed descriptions of more than 430 plants found within 10 miles of London. The best botanical artists of the period, including Sydenham Edwards, James Sowerby and William Kilburn, produced stunning hand-coloured copperplate illustrations for every plant.

The project became a personal obsession so all-consuming that Curtis was driven to resign his prestigious position as Director of the Chelsea Physic Garden. Despite getting financial backing from hundreds of benefactors, whom he named in his book (as we have done ours), he only narrowly escaped bankruptcy thanks to the late intervention of a wealthy patron.

In any case, Curtis wasn't able to sell more than 300 copies of *Flora Londinensis* – two of which Kew is fortunate to have in its library. The book earned him plenty of praise, but no 'pudding', as he called it.

For that, he launched *The Botanical Magazine* in 1787, spotting a gap in the market for a periodical that combined botany and gardening. Again, he championed breathtaking artwork, but this time it was an instant commercial success. Still running as *Curtis's Botanical Magazine*, it's the oldest, continually published (currently by Kew), illustrated botanical periodical in the world.

CONTENTS

INTRODUCTION 9
A NOTE ON USING THIS BOOK 12
THE BASICS OF PLANT ANATOMY 13

EAT

Dandelion, *Taraxacum officinale* 16
Lady's smock, *Cardamine pratensis* 18
White deadnettle, *Lamium album* 20
Pepperwort, *Lepidium campestre* 22
Parasol mushroom, *Macrolepiota procera* 24
Good King Henry, *Blitum bonus-henricus* 26
Earthnut, *Bunium bulbocastanum* 28
Wild rocket, *Diplotaxis tenuifolia* 30
Stinkhorn, *Phallus impudicus* 32
Wood sorrel, *Oxalis acetosella* 34
Lady's bedstraw, *Galium verum* 36
Nettle, *Urtica dioica* 38
Marsh marigold, *Caltha palustris* 40
Wild clary sage, *Salvia verbenaca* 42
Tree oyster mushroom, *Pleurotus ostreatus* 44
Bitter vetch, *Lathyrus linifolius* 46
Jack-in-the-hedge, *Alliaria petiolata* 48
Sow thistle, *Sonchus oleraceus* 50
Corn poppy, *Papaver rhoeas* 52
Oregano, *Origanum vulgare* 54
Mallow, *Malva sylvestris* 56
Meadowsweet, *Filipendula ulmaria* 58

MAKE

Greater periwinkle, *Vinca major* 62
Yellow flag, *Iris pseudacorus* 64
Wild thyme, *Thymus serpyllum* 66
Hedge bindweed, *Calystegia sepium* 68
Soapwort, *Saponaria officinalis* 70
Traveller's joy, *Clematis vitalba* 72
Dog rose, *Rosa canina* 74

GROW

Snake's head fritillary, *Fritillaria meleagris* . . . 78
Cow parsley, *Anthriscus sylvestris* 80
Hart's tongue fern, *Asplenium scolopendrium* . . 82
Yellow rattle, *Rhinanthus minor* 84
Water violet, *Hottonia palustris* 86
Bastard balm, *Melittis melissophyllum* 88
Bee orchid, *Ophrys apifera* 90
Honeysuckle, *Lonicera periclymenum* 92
Houseleek, *Sempervivum tectorum* 94
Toadflax, *Linaria vulgaris* 96
Teasel, *Dipsacus fullonum* 98
Orpine, *Hylotelephium telephium* 100
Scarlet pimpernel, *Lysimachia arvensis* 102
Flowering rush, *Butomus umbellatus* 104
Ivy, *Hedera helix* 106
Swan's-neck thyme-moss, *Mnium hornum* . . 108

KILL

Hemlock, *Conium maculatum* 112
Lily of the valley, *Convallaria majalis* 114
Bulbous buttercup, *Ranunculus bulbosus* . . . 116
Deadly nightshade, *Atropa bella-donna* 118
Pheasant's eye, *Adonis annua* 120
Thorn apple, *Datura stramonium* 122
Green hellebore, *Helleborus viridis* 124
Black nightshade, *Solanum nigrum* 126
Lords-and-ladies, *Arum maculatum* 128
Wood anemone, *Anemonoides nemorosa* 130
Fool's parsley, *Aethusa cynapium* 132

HEAL

Cowslip, *Primula veris* 136
Field scabious, *Knautia arvensis* 138
Broad-leaved dock, *Rumex obtusifolius* 140
Fumitory, *Fumaria officinalis* 142
Common polypody, *Polypodium vulgare* . . . 144
Daisy, *Bellis perennis* 146
German chamomile, *Matricaria chamomilla* . 148
Milk thistle, *Silybum marianum* 150
Wild valerian, *Valeriana officinalis* 152
Rosebay willowherb,
Epilobium angustifolium 154
Milkweed, *Euphorbia peplus* 156
Comfrey, *Symphytum officinale* 158
St John's wort, *Hypericum perforatum* 160
Herb Robert, *Geranium robertianum* 162
Plantain, *Plantago major* 164
Sneezewort, *Achillea ptarmica* 166
Selfheal, *Prunella vulgaris* 168

CONTRIBUTORS 171
INDEX OF PLANTS 172
BENEFACTORS 18TH CENTURY 174
BENEFACTORS 21ST CENTURY 175

INTRODUCTION

The city isn't the first place most people go for plant life. It's probably the last. Town versus country; urban versus rural; concrete versus earth: cities sit at the wrong end of the axis, in opposition to nature somewhere else, out there in the countryside. Or so it seems.

As the great English botanist, William Curtis, knew, that is simply wrong. The sterility of the city is pure illusion. When he published the first part of *Flora Londinensis* in 1775, he set out to document, record and, with the help of three of the most accomplished naturalist artists of the age, illustrate a comprehensive guide to the flora growing in and within 10 miles of London. By the time the last part of his magnum opus came out in 1798, the year before his death, he had described in unparalleled detail more than 430 London plants and demonstrated the astounding diversity on every Londoner's doorstep.

This seminal work is the basis and inspiration for *The Botanical City*, a collaboration between the Royal Botanic Gardens, Kew and Hoxton Mini Press. Reproducing a selection of *Flora Londinensis'* most exquisite plates, accompanied by some of Curtis's choicest insights, we explore and celebrate the most exciting, beautiful, useful and extraordinary plants that you can expect to find on kerbside and roadside, not only in London, but in many temperate cities around the world.

With the benefit of more than 220 years of botanical enquiry since Curtis's death, we also provide up-to-date information and descriptions as well as long-forgotten folklore and herbal traditions. In these pages, you may learn ancient wisdom and benefit from healing powers, enjoy nutritious recipes and new flavours, turn plant fibres into baskets, avoid the deadliest poisoners, and rediscover the essential beauty of plants otherwise taken for granted.

The great botanical books of the past could often be divided between 'floras' – interested in the biology of plants of a particular region – and 'herbals' which focused on plant uses. *Flora Londinensis* was a classic flora, but with a vision and scope extending far beyond standard technical texts and reaching out to a far greater audience than students and botanists. For each plant profile, Curtis combined detailed botanical descriptions of the plants and their habitats with his herbal knowledge as an apothecary, as well as run-downs of the most useful observations from other botanists about the plants' applications and folk history. This, he hoped, would lay a foundation 'for numberless improvements in medicine, agriculture, etc.' for the benefit of all.

Curtis also wanted his book to have broad appeal and be 'useful to the public, as well as instructive and entertaining'. His aim has never been more relevant. It's the guiding impulse behind *The Botanical City*. The energy that draws millions to cities also has a tendency to make people rush more and consider less. In the melee, it's easy to forget that we share our cities with quite a rogues' gallery of plants, the still and silent counterparts to our noise and haste. Whether you call them wildflowers or weeds, every single one of them has something special to say, here and now. If you ever needed it, this book gives you an excuse to slow down, stop and see what's been there all along.

For ease of reference, the plants in *The Botanical City* are divided according to suggested use, between Eat, Make, Grow, Kill and Heal. 'Suggested' because the divisions are arbitrary; many of these plants could sit in several categories, and some of them in all. You can grow any yourself, whether to harvest or admire. Cultivating from seed or garden nursery stock, especially when bought from reputable sources, means you know what you're growing – particularly important if you plan to eat or use them as herbal remedies, which, in any case, is something you should always do judiciously and cautiously.

If he were with us today, Curtis would surely have been delighted that his plants of London are the same now as they were before – growing, seeding, germinating with impressive resilience while all else around them has changed beyond recognition. Clearly, nature's reach doesn't end at the city gates. It bursts through, finding succour in the unusual mini-habitats that only cities provide. The varied microclimates conjured up by the heat of baked concrete and tarmac, the wind and deep shade of towering skyscrapers, the dust of the car-beaten verges, the early industrial canals and waterways, and post-industrial waste grounds create a kaleidoscope of conditions in which a vast array of plants can thrive.

Then there are the parks, green spaces and gardens – including the physic and botanical gardens, such as Kew itself in London – that comprise a substantial chunk of any city. Domestic gardens alone, for example, account for fully a quarter of London's surface. Cities are filled with plants chosen for their use or beauty, for their love of light or shade, for different soils, for ground cover or height, for winter flowers or autumn fruits, engendering a floral cosmopolitanism that's lacking in the field and pasture of the countryside. For that matter, wherever farming is at its most intensive, the pressure on plants and wildlife is often greatest. Many resourceful species have actually found refuge from weed killers and herbicides in the city, where they're often just left to get on with it. If a weed is a plant out of place, then on urban streets all plants are misfits. Here, everything and nothing is a weed.

The fact that urban flora is so versatile, and boasts so many functions, hints at the incredible importance of plant life for humanity. Quite aside from the everyday unseen benefits – not least the ability to temper the heat of the built environment like miraculous air conditioners, absorbing our pollution and giving us oxygen in return – the truth is that plants underpin all life on Earth.

While we may grow, harvest, use and enjoy plants, we should remember that entire ecosystems depend on them too. Each individual plant, even in the city, may help support many dozens of different birds and invertebrates, not just the bees and butterflies that many of us already champion, but slugs and snails, and worms and woodlice in the hidden worlds beneath the pavement, all crucial links in the food chain.

Explore, observe, celebrate – but let wildflowers be. It has been humanity's greatest mistake to think that plants are here purely for our benefit, when in reality we are only here because of the plants.

A NOTE ON
USING THIS BOOK

The plants in this book were recorded in London over 200 years ago but you can still find them in just about any city in the northern temperate zone. That's a vast area between the tropics and the Arctic Circle, where there are four seasons, the winters aren't too cold and the summers not too hot.

Some plants are easier to spot than others – but that's part of the fun of becoming a plant hunter. A chance encounter reminds us that cities are far greener than we think, containing parks, gardens, woods, marshes, rivers and fields, as well as concrete and tarmac. Every plant is part of a bigger story, each supporting all kinds of wildlife both above and below the ground. That's why we should never uproot wild plants, and only pick them in moderation where the law allows.

Eating and even touching wild plants can be dangerous. This book celebrates city weeds and wildflowers, but it's not a medical manual, cookbook or identification guide. We encourage you to grow your own plants at home and to seek expert advice before using any wild plants for any purpose.

THE BASICS OF PLANT ANATOMY

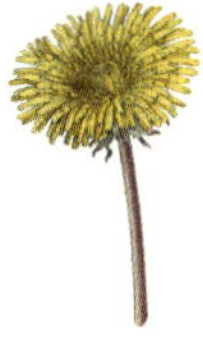

FLOWERS — Flowers produce seeds to make new plants. They must be pollinated first. Grains of pollen from male parts fertilise female parts either in the same or different flowers. Some plants can pollinate themselves, while others rely on insects, birds and mammals or the wind to carry pollen for them.

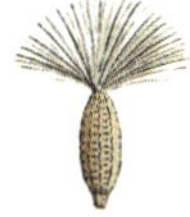

SEEDS — Seeds form in pollinated flowers, often inside protective fruits, and are the next generation of the plant. They can be spread by the wind, insects and animals or just fall to the ground, where in the right conditions they germinate and grow into new plants.

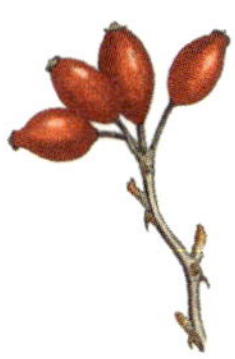

FRUITS — Many flowering plants produce fruits holding seeds. Fruits are often sweet and juicy to encourage birds and animals to eat them and disperse the seeds.

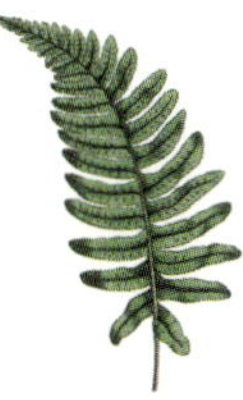

LEAVES — Leaves create food for the plant through photosynthesis, a process whereby the sunlight is absorbed and turned into sugars that the plant uses for energy.

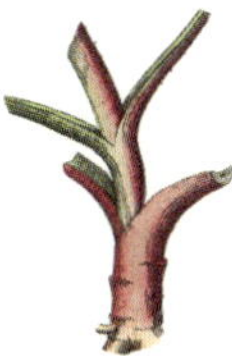

STEMS — The stem connects the roots to the other parts of the plant, usually above the ground. The stem produces leaves and flower buds and gives the plant structure as well as containing the transport system for the plant, moving food and water from roots to leaves and flowers.

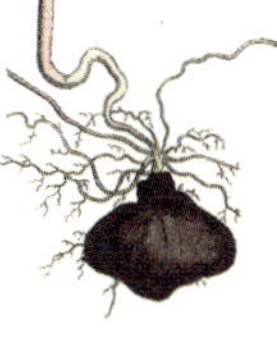

TUBERS — Tubers are adapted stems and roots that swell mainly to store starch, which the plant can process into sugars when needed.

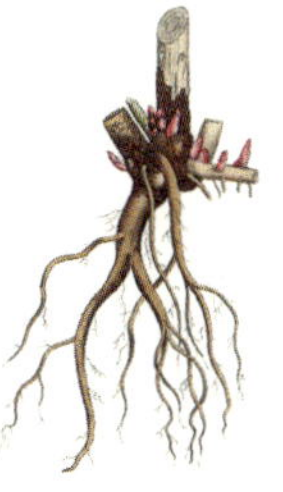

ROOTS — Roots usually anchor the plant in the ground and draw water and nutrients from the soil.

RHIZOMES — Rhizomes are specialised stems that grow horizontally along or just under the ground. They store nutrients for when food is scarce and can produce roots and shoots to make new plants.

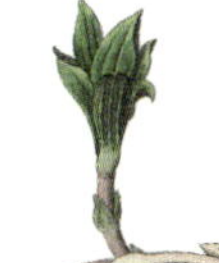

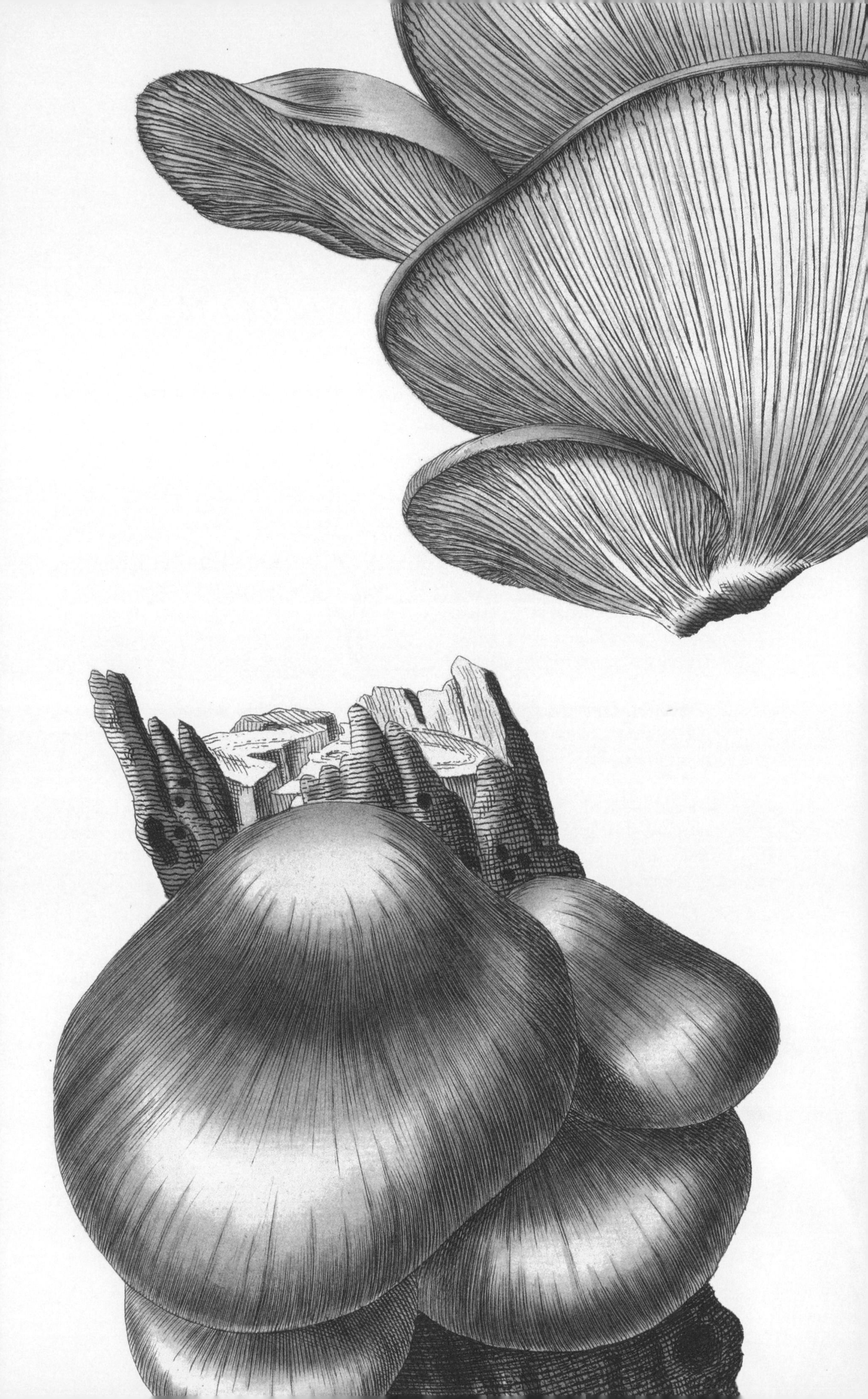

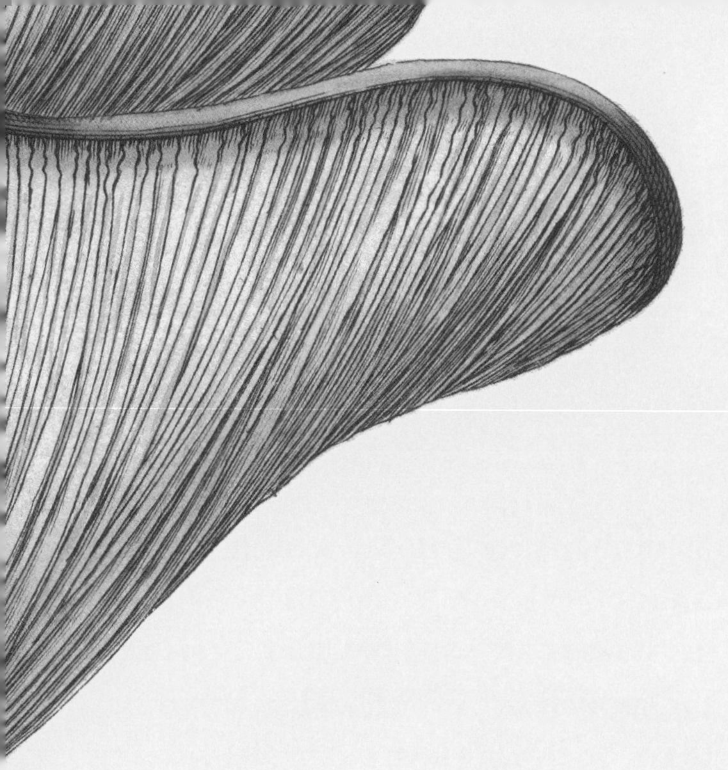

EAT

Fresh, free and *ALWAYS SEASONAL*
these plants and fungi are the most delicious
and *NUTRITIOUS* in nature's
store cupboard and have the smallest
carbon footprint of any food around.

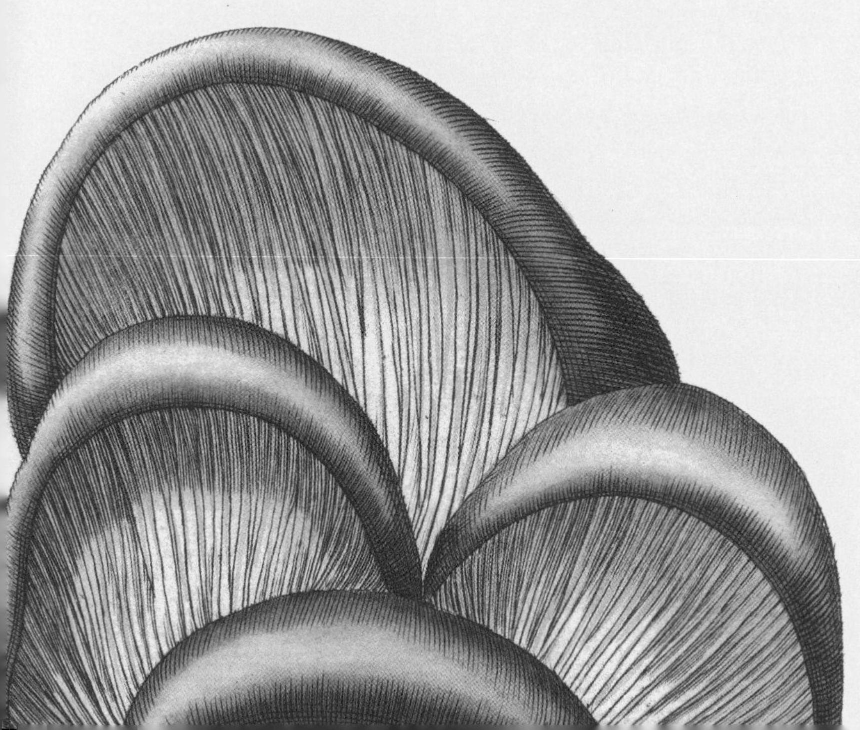

DANDELION

Taraxacum officinale

The humble dandelion, with its gaudy sunbursts of yellow florets, does more to cheer up our urban landscape than any other plant. It's not just everywhere, but it's edible from tip to taproot. Its flowers are used in wines, beers and soft drinks, its roots can be roasted and ground like coffee, and its leaves and stems burst with vitamins and minerals. Who doesn't love puffing its pleasing spherical seed heads, every child's favourite disposable timepieces, to the wind, ensuring the plant stays as plentiful as it is bountiful?

ANATOMY NOTES

FLOWERS — Flowerheads, made of hundreds of petal-like florets, each a flower in its own right, are used in infusions, drinks, skin balms, soaps, syrups and jellies.

SEEDS — The fluffy 'pappus' (from the Greek for 'grandfather', because it looks like a grey beard) opens when conditions are best for long-distance travel on the breeze.

LEAVES — Jagged like the teeth of a lion (*dents de lion* in French, which is how we get 'dandelion'), the leaves can be sautéed, boiled, or simply eaten raw. They're loaded with vitamins A, C and K, iron, potassium, manganese and calcium. Deprive them of sunlight for two weeks (blanching) for milder, sweeter leaves.

STEMS — The white goo that oozes from a broken stem is a latex that can be made commercially into rubber products, even car tyres.

ROOTS — They're best harvested in autumn for medicinal use, but spring for the kitchen when fructose levels are higher. Cut into strips and air-dry before using in infusions, tinctures or coffee. They're used to improve digestion and as a diuretic, which explains its colourful common name, 'piss-a-bed'.

ESSENTIALS

Aka — Piss-a-bed, blowball

Family — Asteraceae (daisy)

Urban habitat — Widespread in grass, disturbed and cultivated areas

Height — 35 cm / 1 ft 2 in

Flowers — May to October

Uses — Food, diuretic, rubber

Grow — Sow seed in spring

Harvest — Flowers in summer, leaves in early spring, roots in spring or autumn

BOTANIST'S RECIPE

SALAD — Wash and dry 350 g / 12 oz of dandelion leaves. While you boil 5 eggs, brown 5 chopped rashers of smoked streaky bacon in a hot pan. Add the bacon to the leaves. Pour 5 tablespoons of cider vinegar into the pan and scrape with the bacon juices with a wooden spoon, reducing a little. Peel and slice the eggs, add to the salad, and coat with the vinegar glaze. Enjoy with a few twists of fresh pepper and a glass of red wine.

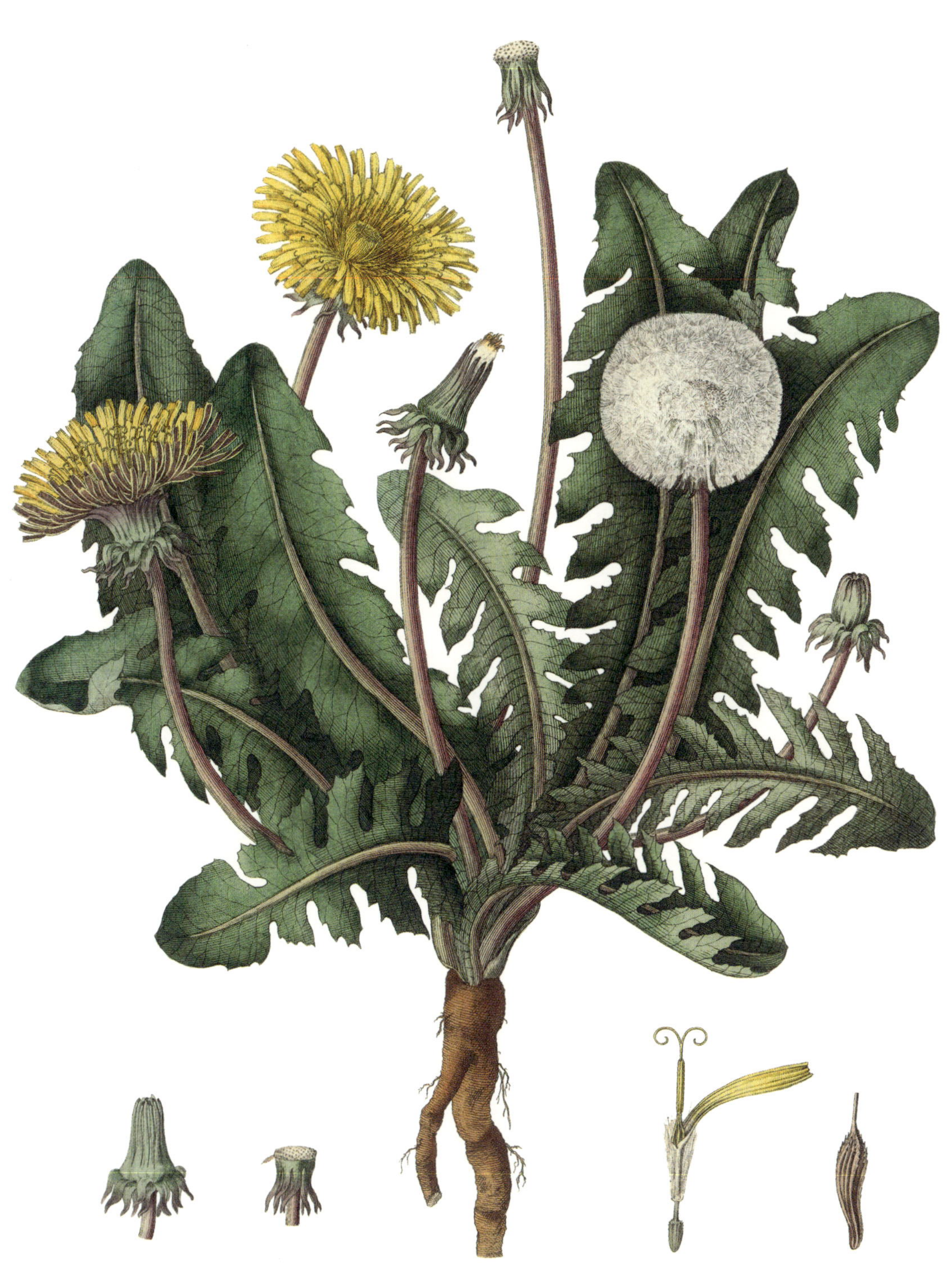

Taraxacum officinale

LADY'S SMOCK

Cardamine pratensis

Just as the grass becomes thick and luscious, and the first calls of the cuckoo ring out, the elegant lilac-veined flowers of lady's smock emerge. Spring must be here. In fact, the plant's appearance coincides so closely to the cuckoo's springtime arrival from Africa that the plant has earned the nickname 'cuckoo flower'. More than simply a charming meadow wildflower, the leaves, shoots and flowers of lady's smock are also edible, with a spicy mustard kick that only intensifies the older the plant is. Pick young leaves to add to salads; the flowers are milder and glorify even the humblest springtime dish.

ANATOMY NOTES

LEAVES — Not only nutritious, packed as they are with vitamin C, they're also supposed to help with eczema and arthritic pain. You can either eat them or make them into tea for the purpose. The leaves produce their own roots when they touch damp soil and eventually form new plants.

FLOWERS — Lady's smock is one of the main food sources for the orange-tip butterfly, so much so that it takes its scientific name *Anthocharis cardamines* from the plant. The flowers are their favourite part, but they also eat the leaves, stems and seeds. Folklore has it that adders are also attracted to the flowers, which is why it was considered bad luck to pick and take them indoors. Do so and you'll be bitten within a year, apparently!

BOTANIST'S RECIPE

SOUP — For a spicy soup to warm you up on a fresh spring day, peel and chop 2 onions, 2 potatoes and 2 garlic cloves. Fry the onions in a pan until soft, then add the garlic and potatoes. Pour in 400 ml / 14 fl oz of vegetable stock and simmer until the potatoes are soft. Add 3 big handfuls of washed and chopped lady's smock leaves and cook for a further 3 minutes. Blend until smooth, season, and serve with a swirl of crème fraîche, a scattering of the flowers, and a big chunk of warm bread.

ESSENTIALS

Aka — Cuckoo flower

Family — Brassicaceae (cabbage)

Urban habitat — Damp grassy areas, watersides

Height — 40 cm / 1 ft 4 in

Flowers — April to June

Uses — Food, ornamental

Grow — Sow seed in spring

Harvest — Leaves and shoots in early spring, flowers in spring

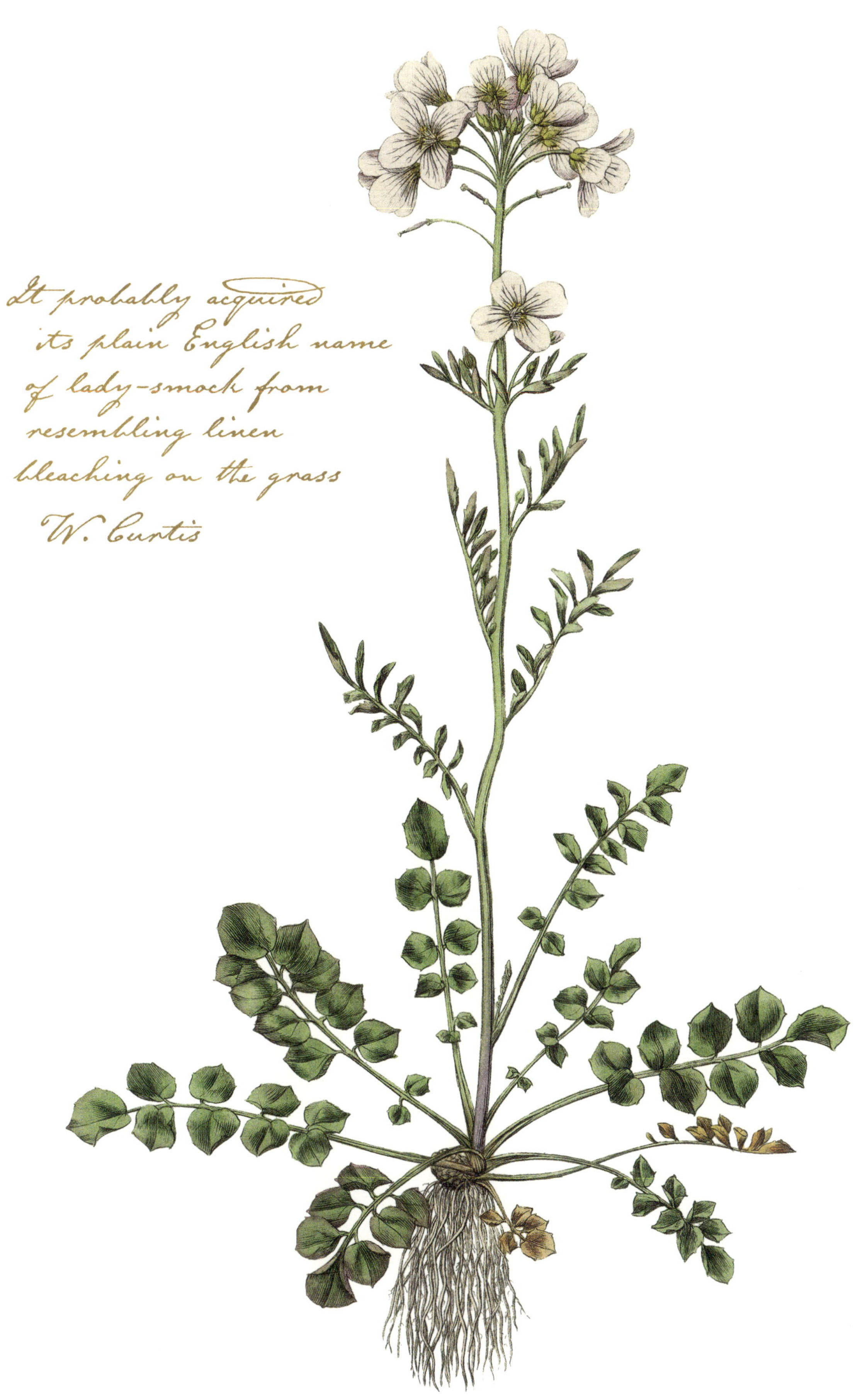

Cardamine pratensis

WHITE DEADNETTLE

Lamium album

White deadnettle is the great pretender. With little to defend itself, it mimics the toughest plant on the block, the stinging nettle – one of the best bodyguards around. Before its flowers appear and reveal its identity, you have to be brave or eagle-eyed to dare touch it, but your skill and courage will be rewarded. Its tender young leaves and stem tops are actually edible and crammed with nourishing vitamins and minerals (especially vitamin A), and its blanched flowers make a very pretty garnish or restorative cup of tea.

ANATOMY NOTES

LEAVES — Not just great in salads and omelettes, the leaves can be made into a medicinal tea to relieve coughs and sore throats. They have anti-inflammatory properties too. Their rich green pigment was very helpful to the Russian-Italian botanist, Mikhail Tsvet, in his pioneering experiments in the invention of chromatography – a technique to separate of chemical mixtures.

FLOWERS — This plant is also pretending with its flowers: they appear to be whorls, radiating from a central point, but are actually 'false whorls' arranged in a pattern known as a verticillaster. The fluted base of the flower resembles a long white throat, which gives the plant its scientific name (*lamos* is Greek for throat, and *album* is Latin for white). Each is filled with nectar and pollen, accessible to long-tongued insects such as bumblebees, which throng to it.

BOTANIST'S RECIPE

SALAD — White deadnettle leaves make a handsome salad on their own, but they're especially satisfying when harmonised with a scattering of other wild greens, such as dandelion, wild rocket and wood sorrel. You could bulk it out with olives, fennel seeds, tomatoes and cucumber, plus nasturtium, calendula or borage flowers. A very light dressing works best, such as olive oil and lemon juice – and why not pour yourself a glass of meadowsweet champagne (see p.58) for good measure?

ESSENTIALS

Aka — White archangel

Family — Lamiaceae (mint)

Urban habitat — Waysides and wasteland, often near nettles

Height — 50 cm / 1 ft 8 in

Flowers — May to December

Uses — Edible leaves and flowers

Grow — Sow seed in spring

Harvest — Leaves in late spring, flowers in the summer

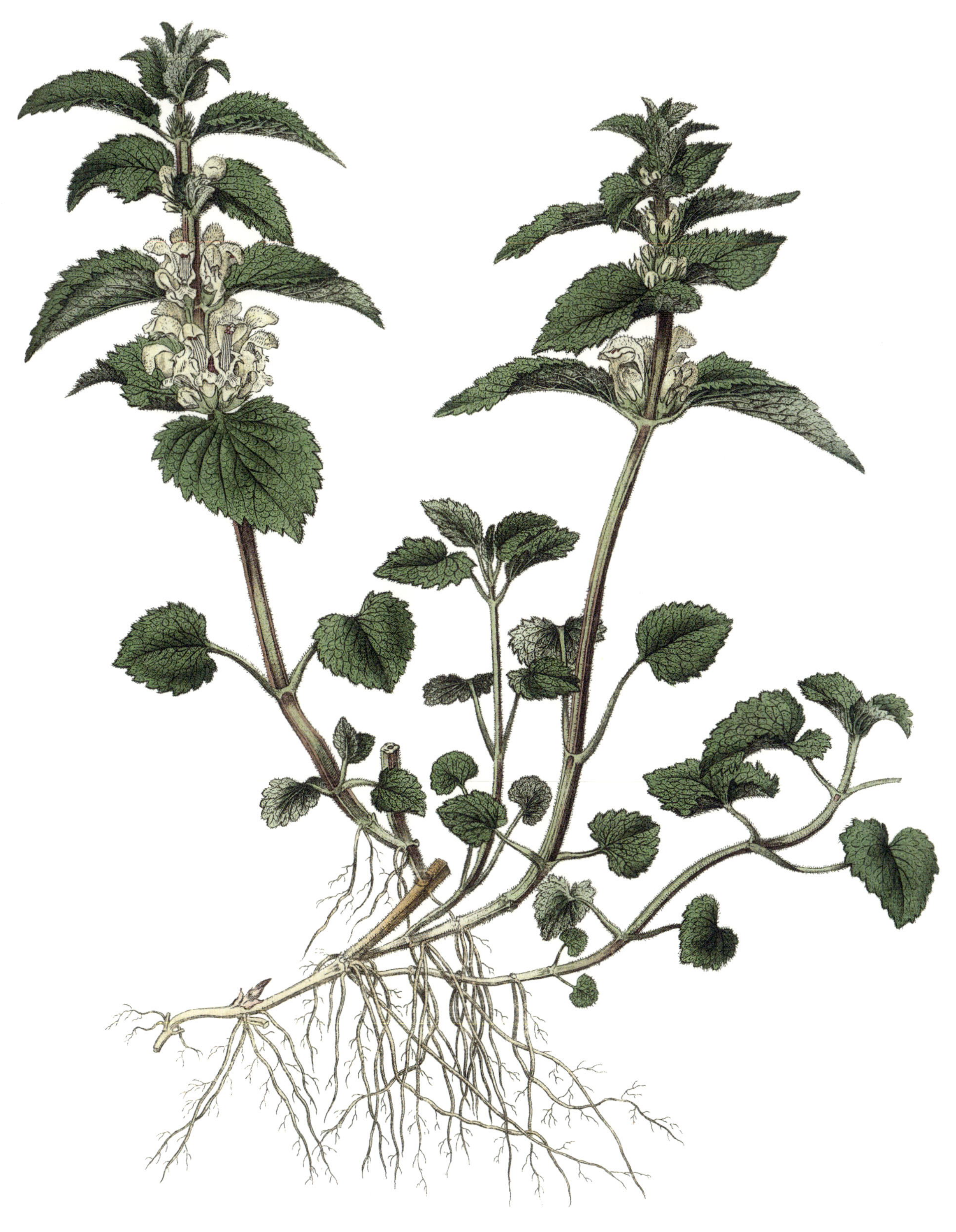

Lamium album

PEPPERWORT

Lepidium campestre

What it may lack in the looks department, pepperwort more than makes up for in flavour. Yes, it's peppery, and its leaves and shoots can pep up any recipe in place of spinach or watercress. You can also grind its dried seeds like peppercorns. And it deserves a special commendation for sheer tenacity. It can grow just about anywhere, sending up throngs of stiff white-tipped spires between broken concrete slabs or from fissures in a brick wall, like a city skyline in miniature.

ANATOMY NOTES

LEAVES — The leaves are full of vitamins A and C, and proteins, making them a highly nutritious ingredient. In traditional herbalism, it was used to treat for stomach complaints, skin disorders and irregular heartbeats.

SEEDS — The seeds are rich in oils good enough to be used in machinery and vehicles. The plant's fast and high-yield lifecycle means farmers can grow it between other crop harvests and in colder climates that aren't usually very productive for seed oils.

SEEDPODS — The seedpods, full of seeds, can be eaten raw, and are a spicy, crunchy snack to have with a cool drink on a summer's day.

BOTANIST'S RECIPE

MUFFINS — Get your day off to a flying start with these energising breakfast muffins. Whisk 2 eggs, then stir in 150 ml / 5 fl oz milk and 75 g / 3 oz melted butter. Mix in 150 g / 5 oz grated mature cheddar cheese, 75 g / 3 oz washed pepperwort leaves, and 250 g / 9 oz self-raising flour. Season with salt and pepper and crumble in half a vegetable stock cube, then pour into a greased muffin tray. There should be enough for 12 muffins. Bake in a pre-heated oven at 180 °C / 355 °F (160 °C fan / 320 °F) for 20 to 25 minutes.

ESSENTIALS

Aka — Bastard cress

Family — Brassicaceae (cabbage)

Urban habitat — Disturbed ground, cracks in the pavement

Height — 40 cm / 1 ft 4 in

Flowers — May to August

Uses — Edible leaves, shoots and seeds

Grow — Sow seed in spring

Harvest — Leaves in spring, seed in autumn

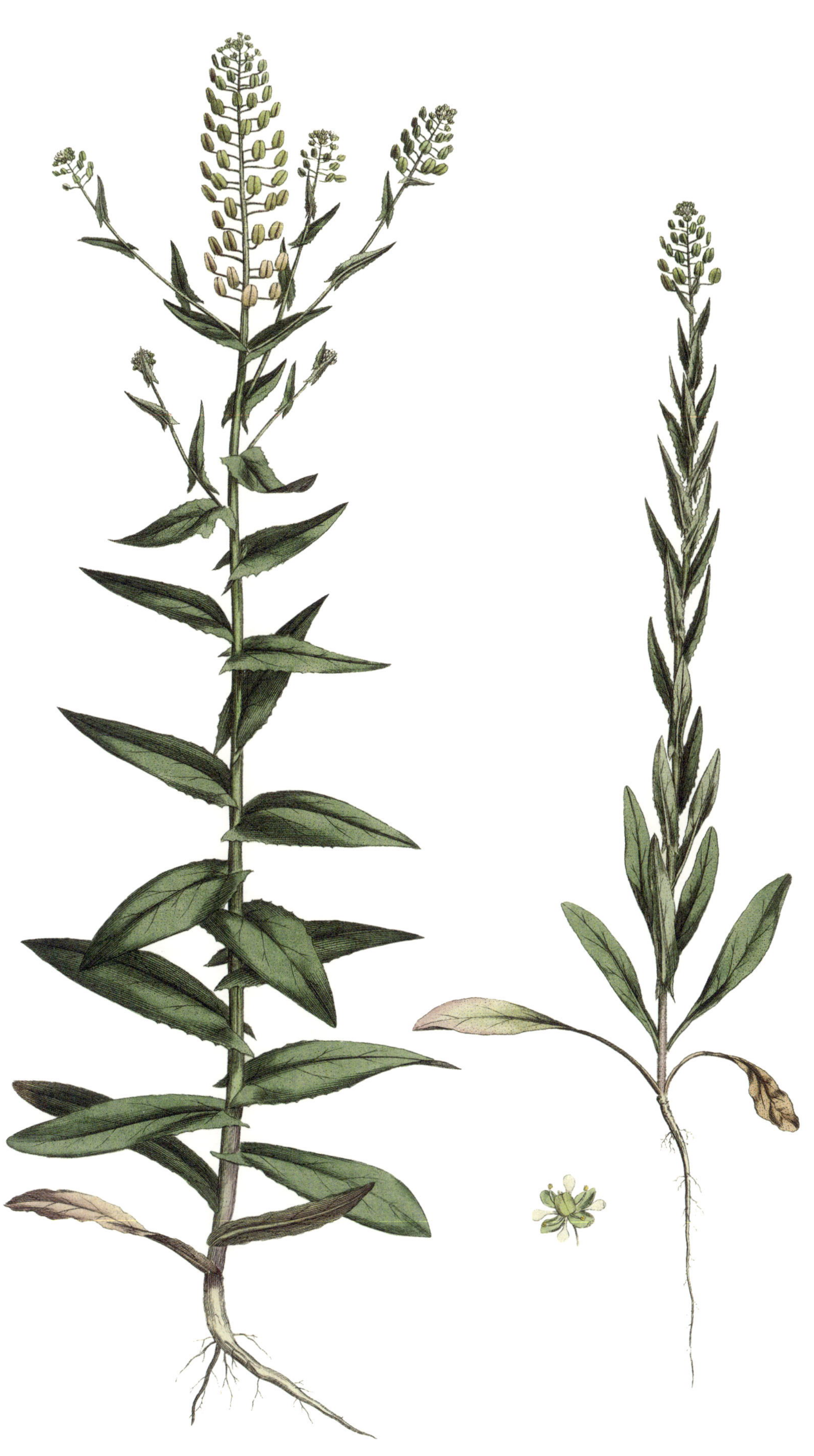

Lepidium campestre

PARASOL MUSHROOM

Macrolepiota procera

As a parasol, this magnificent mushroom could provide shade for not one, but two gnomes, and probably a few of their fairy friends to boot. They come bigger than your dinner plate – which is just where you'll want to put them because they're so scrumptiously good to eat. Whether fried in oodles of butter with garlic, coated in breadcrumbs like a schnitzel, or baked, stuffed with bacon for breakfast with scrambled eggs (in full Slovakian style), a feast awaits. And they could be lurking in a lush roadside verge near you, if the gnomes don't get them first.

ANATOMY NOTES

CAP — The cap starts out bulbous and flattens, extending to a final diameter of about 30 to 40 cm / 12 to 16 in. Its flaky rippled pattern appears as its skin stretches. The cap is the only part of the mushroom that's edible. As always with fungi, be sure you know what you're eating.

MYCELIUM — The mycelium, which is a bit like the roots of a fungus, extracts nutrients and water from the soil. It can form a mutually beneficial relationship with ants, providing material to help the insects strengthen their nests, who go on to spread the mycelium to make new mushroom colonies.

STIPE — The stipe (stalk) is hollow and very fibrous, so it's no good for a meal. Its surface is a little scaly like snakeskin, hence one of its common names, snake's hat.

BOTANIST'S RECIPE

SLOVAKIAN BREAKFAST — Wash and dry your mushroom tops. Stuff the smaller ones with strips of bacon, folded up like a concertina. Season with salt and pepper, sprinkle with paprika and olive oil, and place in an oven heated to 200 °C / 390 °F (180 °C / 355 °F fan) for 20 minutes, turning once. Slice your larger mushrooms. Chop and fry some bacon and onion for 5 minutes, then add the sliced mushrooms and some caraway seeds. Cover and simmer, adding a little stock if necessary, until softened. Add beaten eggs and scramble. Serve with your baked mushrooms.

ESSENTIALS

Aka — Tall mushroom, snake's hat

Family — Agaricaceae

Urban habitat — Shady woods, verges, lawns and grassy margins

Height — 40 cm / 1 ft 4 in

Uses — Food

Grow — Buy spores to grow your own

Harvest — July to November

This mushroom, inferior to few in point of elegance, is frequently found in woods, and dry hilly pasture

W. Curtis

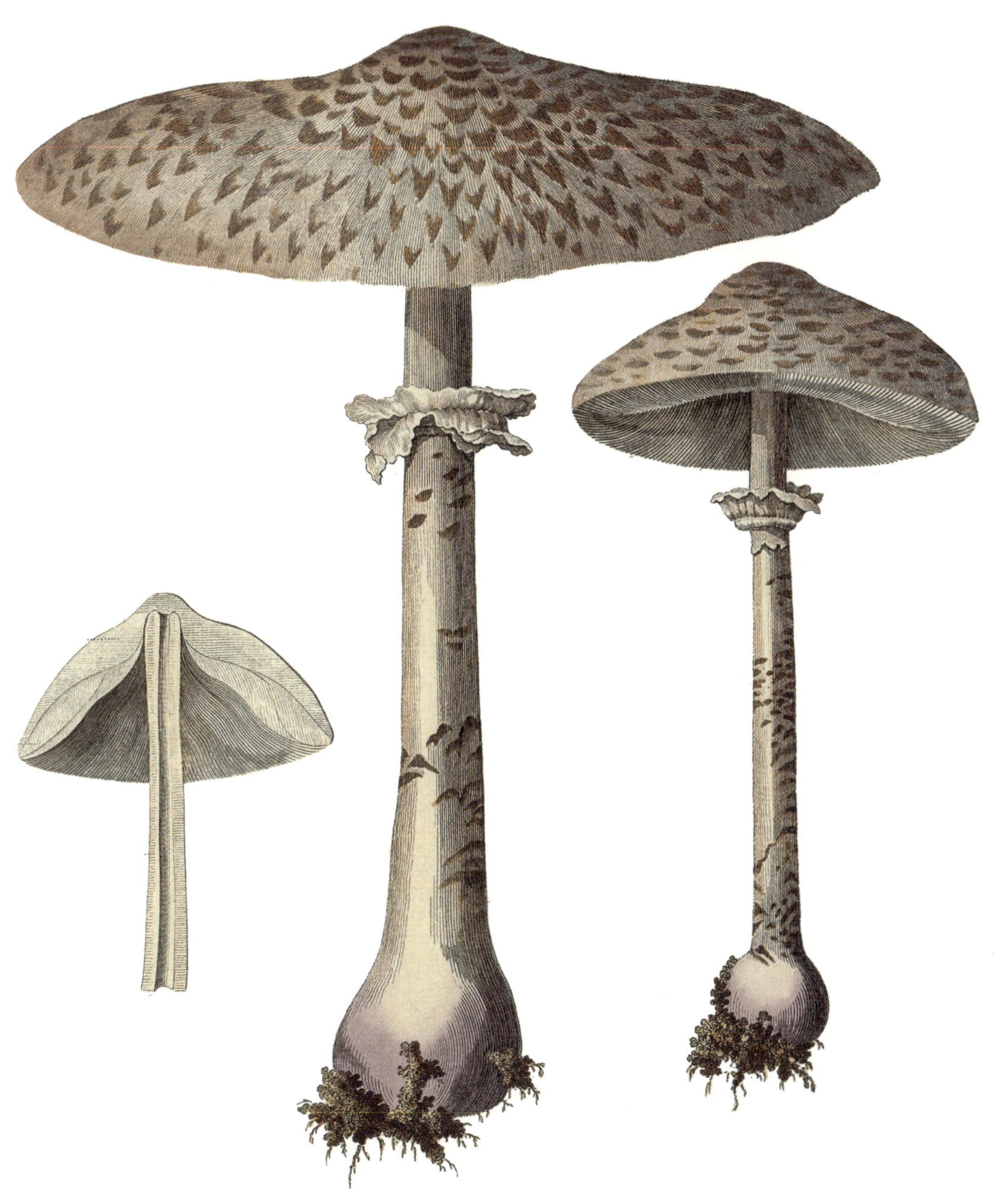

Macrolepiota procera

GOOD KING HENRY

Blitum bonus-henricus

An easier crop plant to have in your garden is hard to come by. Once established, Good King Henry pretty much looks after itself year after year, untroubled by pests or diseases, repaying your neglect of it with armfuls of food – spinachy leaves, asparagus-esque stems, broccolic flower buds, quinoa-like seeds that can be ground into flour. Were it not for the bitterness of the leaves, perhaps it would be on supermarket shelves instead of spinach; in Lincolnshire where it has long been cultivated, it's still preferred by some. Either way, it's due a renaissance. Long live Good King Henry!

ANATOMY NOTES

LEAVES — The earlier you crop the leaves, the less bitter they are. Soaking them in brine also draws out the bitterness before cooking (but don't cook them in the same water). The leaves resemble a goose's foot. In German folklore, 'Heinzl men' are nocturnal sprites sometimes depicted with flat feet like a goose. 'Heinzl' is related to 'Henry' – a possible source for this plant's unusual name.

SHOOTS — Young shoots can be trimmed and eaten like asparagus. If you place a pot over them to keep the sun off for a week before harvesting, they'll be sweeter.

ROOTS — Are your sheep coughing? Have your cows and their milk been cursed again? Both afflictions have traditionally been cured by feeding them the roots of Good King Henry.

BOTANIST'S RECIPE

BREAD — Good King Henrify your next loaf by kneading chopped and blanched young leaves into the dough. For a smoother finish, purée the wilted leaves in a blender. You could even take it to the next level by using a bit of your own GKH flour, made from ground seeds. Remove the husks before you grind – more easily done outside on a windy day when the chaff can blow away. The more you use, the darker your loaf will be.

ESSENTIALS

Aka — Poor-man's asparagus, mercury, Lincolnshire spinach

Family — Amaranthaceae (amaranth)

Urban habitat — Verges and hedges

Height — 40 cm / 1 ft 4 in

Flowers — April to July

Uses — Edible leaves, stems, flower buds and seeds

Grow — Plant seeds in spring

Harvest — Leaves, shoots and flower buds in spring, seeds in autumn

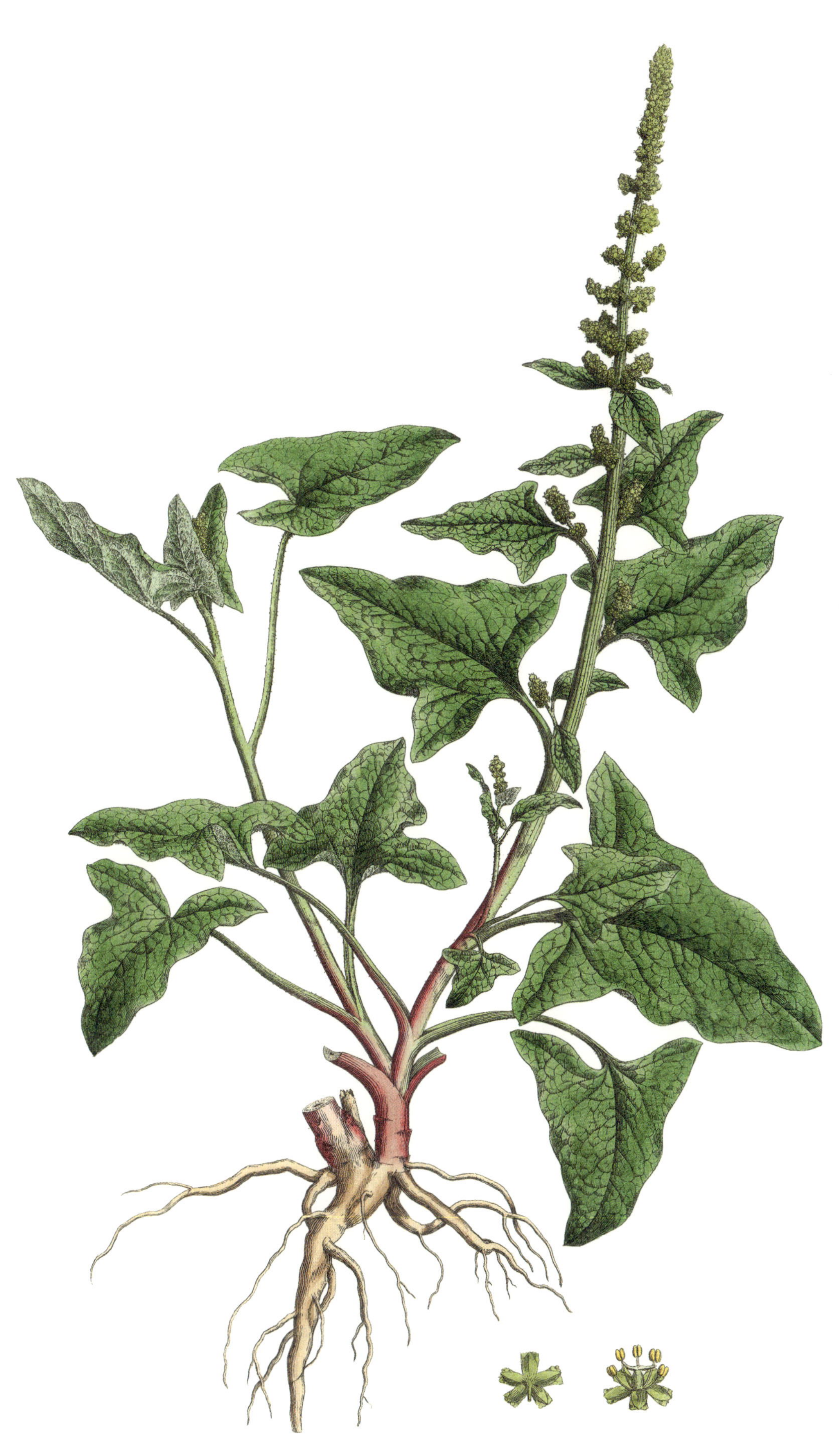

Blitum bonus-henricus

EARTHNUT

Bunium bulbocastanum

The earthnut is a pretty unassuming plant until it bursts into swathes of white flowerheads in early summer, brightening roadsides and verges about town. Autumn's its true moment of glory, though, when it yields a knobbly edible root that both looks and tastes like a chestnut. The tough bit is finding and digging them up. If you happen to have a pig to hand, it will seek them out for you in a trice! They love them; it's not also called a pignut for nothing. Its leaves and seeds are also fit for the kitchen table – but do take care you're not confusing them with its deadly cousin, hemlock (see p.112).

ANATOMY NOTES

LEAVES — The feathery leaves, not a million miles from fennel in appearance, are best gathered sparingly in early spring when they are young and tender, and the plant is still emerging, long before it flowers. The fresh parsley flavour works well mixed into spring salads.

SEEDS — The seeds ripen by early autumn and can be collected for drying at home. Sometimes called black cumin, these spicy seeds frequently appear in Indian subcontinental cuisines. Confusingly, nigella seeds are also known as black cumin.

ROOTS — The tubers can be eaten raw or cooked, and taste something like an earthy chestnut. *Bulbocastanum* translates as 'chestnut-bulbed'.

BOTANIST'S RECIPE

BLACK CUMIN (EARTHNUT) SALT — You can use this to season almost any dish, but it's also great on its own with quails' eggs. Harvest the seeds and air dry thoroughly for around a week. Toast in a dry pan on a medium heat, stirring constantly for two minutes. Grind into powder with a pestle and mortar. Mix one teaspoon of this powder, also referred to as black cumin, to every three teaspoons of good quality sea salt. Add chilli flakes for a kick.

ESSENTIALS

Aka — Pignut

Family — Apiaceae (carrot)

Urban habitat — Roadsides, grassy margins, scrub, hedges

Height — 60 cm / 2 ft

Flowers — May to July

Uses — Food

Grow — Sow seed in autumn

Harvest — Leaves in spring, roots and seeds in autumn

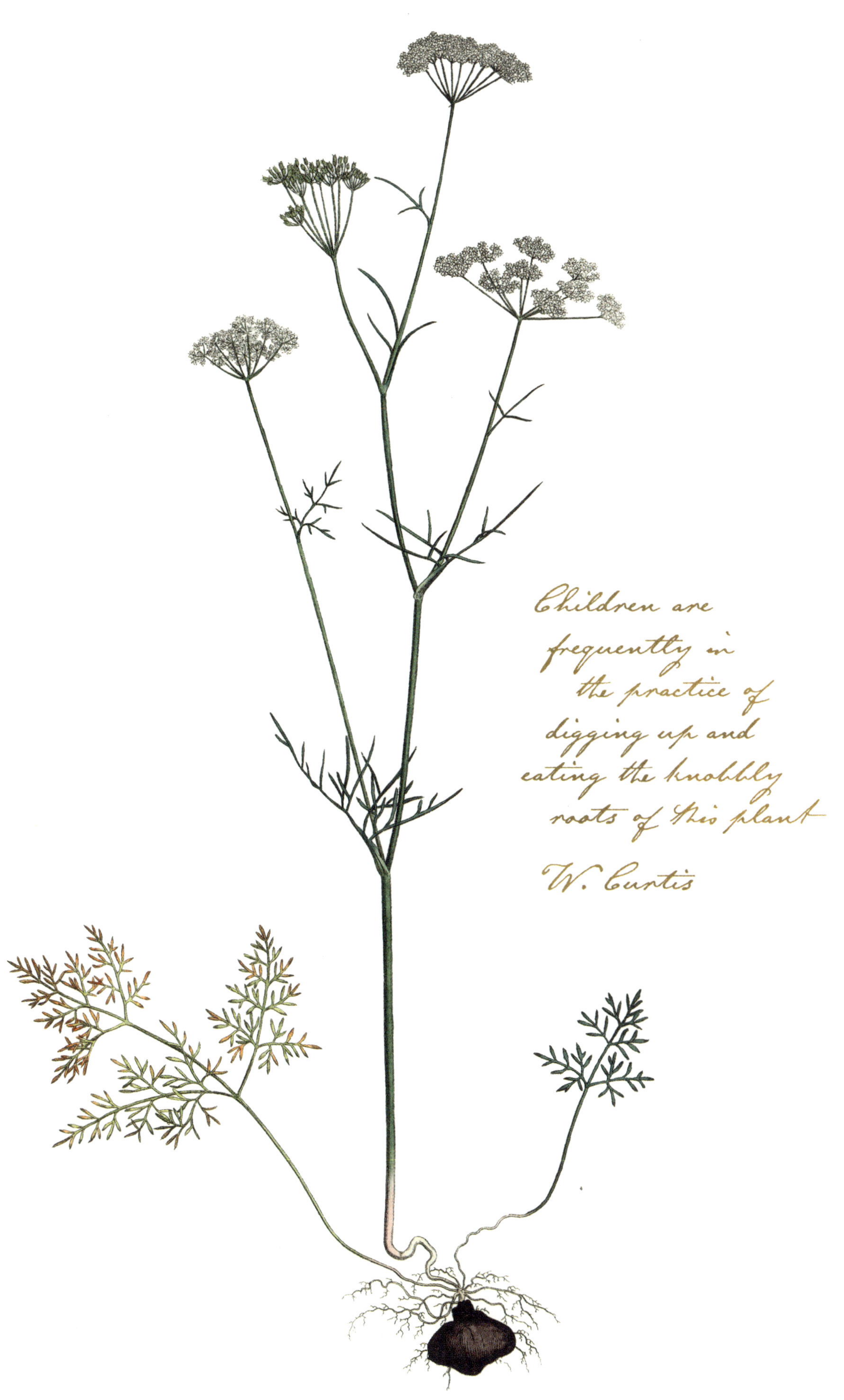

Bunium bulbocastanum

WILD ROCKET

Diplotaxis tenuifolia

Wild rocket is the intense and unruly cousin of its cultivated cousin, our familiar salad bowl essential. Not only more potent in flavour, this untamed relative also packs more of a punch nutritionally, being particularly rich in the antioxidants, vitamin C and quercetin. Growing in abundance in walls and verges, you can quickly forage enough for a sandwich, soup or salad. The leaves are most delicious when they're young and tender, but if you want some oral fireworks, they get spicier with age.

ANATOMY NOTES

SEEDPODS — The seeds and seedpods are edible, and give a crunchy, peppery bite to a salad. You can also grow them as 'microherbs' on the windowsill and harvest them early, usually from about a week onwards.

FLOWERS — Wild rocket's delicate yellow, cross-shaped flowers are beautiful, and piquant, adornments for salads and other dishes.

ROOTS — A long taproot anchors the plant to vertiginous surfaces.

LEAVES — They may be a little spindlier than commercially grown rocket (the Latin *tenuifolia* means 'thin leaf'), but once cut, the leaves will grow back and give around two further flushes of deliciously tender salad leaves, before becoming very bitter as the plant finally runs to seed.

BOTANIST'S RECIPE

PESTO — This is a feisty pesto to enliven pasta, pizza, salads - even a succulent salmon en croûte. Combine 50 g / 2 oz pine nuts, 100 g / 4 oz wild rocket leaves, 50 g / 2 oz parmesan, 150 ml / 5 oz of olive oil and 1 garlic clove in a blender. Whizz into a paste and season. It lasts in the fridge for up to 5 days, or longer if you keep it under a layer of olive oil in a jar.

ESSENTIALS

Aka — Perennial wall rocket, wild arugula

Family — Brassicaceae (cabbage)

Urban habitat — Sunny spots, especially in cracks in walls and pavements

Height — 40 cm / 1 ft 4 in

Flowers — Summer

Uses — Cooking

Grow — Sow in spring

Harvest — Leaves in spring, flowers in summer, seeds in autumn

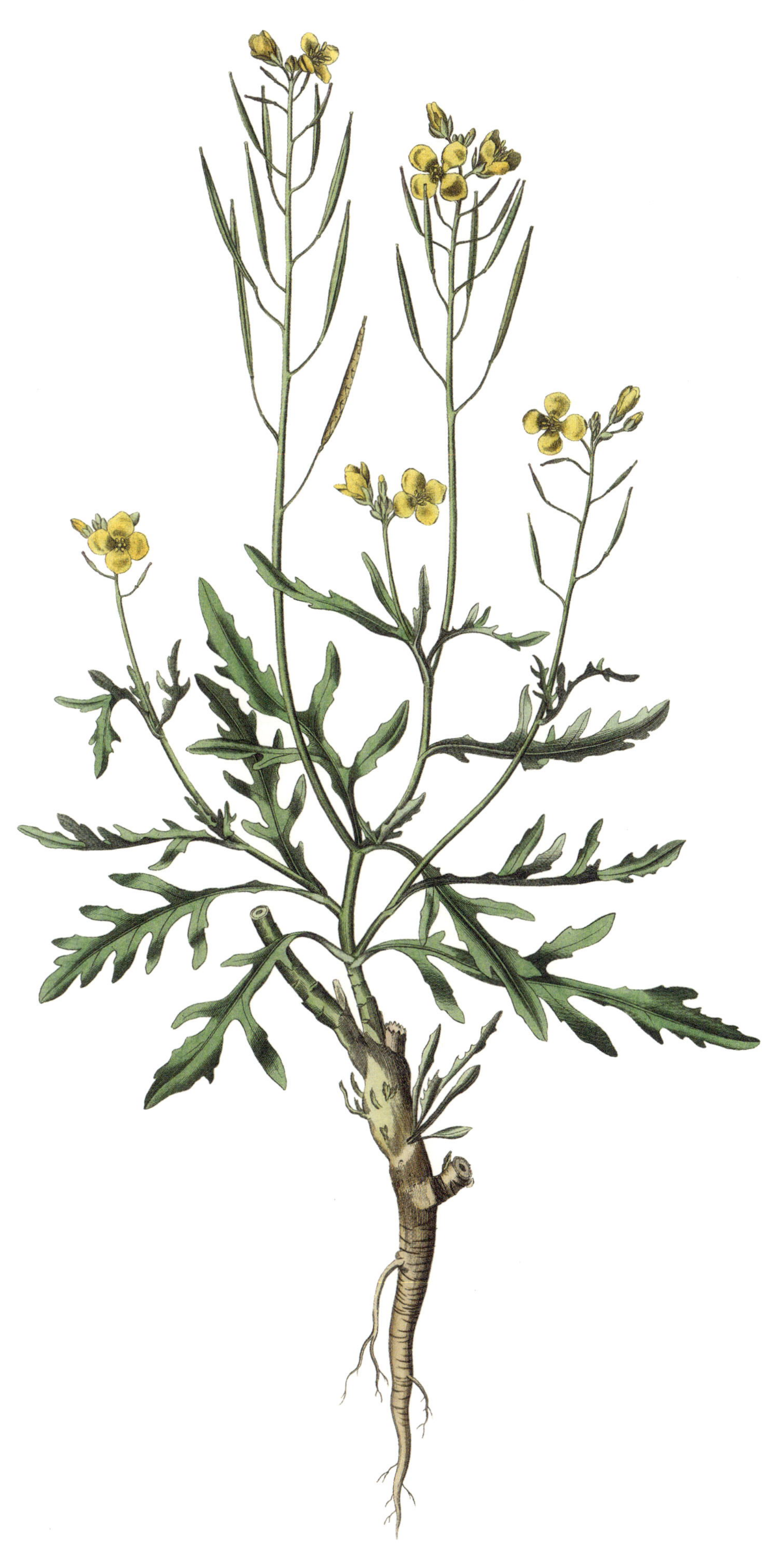

Diplotaxis tenuifolia

STINKHORN

Phallus impudicus

Immodestly phallic (Latin *impudicus* means immodest, after all), this crudely suggestive fungus can crop up wherever there's decaying wood. They're so visually shocking that Victorian zealots would sweep these mushroom members from woodland walks before they could corrupt impressionable minds. These days, you're more likely to be offended by the stench it gives off to attract flies, a putrid simulation of rotting flesh and sewage. It's as unappetising as it sounds. Before the stink and horn appear, however, when the fungus is just an 'egg' in the ground, you have yourself a delicacy that few have tried.

ANATOMY NOTES

EGG — The egg, sometimes called the witch's egg, is the immature fruit. It's good to eat before it sends up the stinking phallus. The earlier you find them, in shallow earth or beneath a thin layer of leaves or pine needles, the less pungent they're likely to be. Be careful not to confuse them with eggs of poisonous *Amanita* mushrooms, which look completely different when cut in half. The stinkhorn has jelly and spores inside, which will turn into the gleba.

STIPE — The stipe (stalk) can pop up out of the egg to its full height in just a few hours. This and other parts of some stinkhorn varieties are considered aphrodisiacs. Whether that's because of their chemistry, how they look, or the alcohol commonly added to love potions is up for debate.

GLEBA — This is the darker, slimy, smelly bit at the top of the stinkhorn where the spores are. Flies come and feast, taking the spores with them to other stinkhorns.

VOLVA — The membrane around the egg is usually discarded before eating.

BOTANIST'S RECIPE

EGGS — Once you've harvested your stinkhorn eggs, cut them open. If you're feeling brave and the egg is young enough you can eat the jelly and spores, which some people like for their intriguing textures. It's the central white bit, the embryonic stalk, that most go for. Crunchy and with a gentle earthy flavour, it can be sliced and eaten raw or cooked.

ESSENTIALS

Aka — Stinking morel, common stinkhorn

Family — Phallaceae (stinkhorn)

Urban habitat — Woods

Height — 25 cm / 10 in

Fruits — Late summer to autumn

Uses — Forest food, aphrodisiac, anticoagulant

Grow — On well-mulched beds

Harvest — Summer and autumn

The fetor arising from it quickly pervading every part of the house, we were obliged to get rid of it

W. Curtis

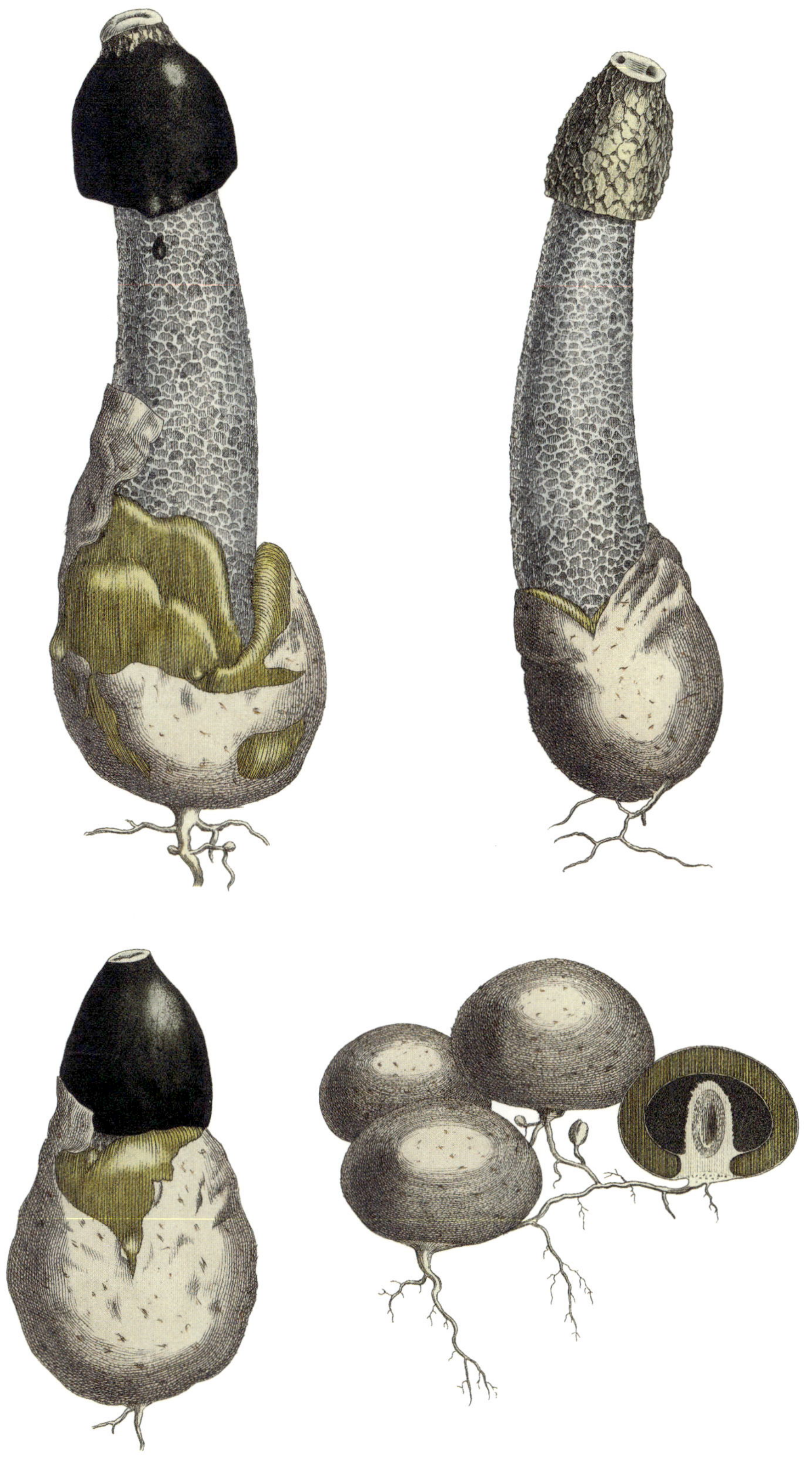

Phallus impudicus

WOOD SORREL

Oxalis acetosella

In the city's quieter, shadier corners you may chance across lime-green leaves shaped like shamrocks, delicately topped with winsome pink-veined flowers. Wood sorrel couldn't look any sweeter. In reality it's filled with oxalic acid, so much so it gave the acid its name. Despite this, wood sorrel is actually great to eat in small quantities for its tangy citrus flavours; only a few leaves are needed to transform a spring salad. Its flowers are also edible – but far too pretty to eat.

ANATOMY NOTES

LEAVES — The leaves forecast the weather. Just before it starts to rain, they fold up and stay that way until it dries up. They also close at bedtime, along with the flowers. They contain oxalic acid which is toxic – but you'd have to eat a lot for problems to appear; many common foods like spinach and rhubarb also have oxalic acid. Herbalists have used the leaves to treat fevers and stomach aches, and research is ongoing on their use for kidney stones and joint pain. Also useful in the house, its essential salt was used to remove rust stains from linen. Its curious common name, cuckoo's meat, may come from the flowers' appearance at about the same time the cuckoo arrives, or from the myth that the cuckoo, the symbol of Hera, goddess of marriage, had to eat wood sorrel to find her voice.

SEEDS — The seedpods burst open on hot days, shooting seeds out at distances of up to 2.5 m / 8 ft 4 in, helping it spread quickly and efficiently – not bad for a 30 cm / 1 ft plant.

FLOWERS — Towards the end of the summer, the flowers do not open, and instead self-pollinate while in bud formation.

ESSENTIALS

Aka — Cuckoo's meat

Family — Oxalidaceae (wood sorrel)

Urban habitat — Shady woods and hedges

Height — 30 cm / 1 ft

Flowers — April to May

Uses — Food

Grow — Sow seed in spring

Harvest — Leaves in spring

BOTANIST'S RECIPE

SALMON AND WOOD SORREL SAUCE — With its citrus taste, wood sorrel goes very well with fish. Cut 750 g / 27 oz salmon fillet into strips and place on a baking tray. Simmer 300 ml / 10 fl oz fish stock, 45 ml / 2 fl oz double cream and 25 ml / 1 fl oz dry vermouth for around 15 minutes to reduce it. Then add another 45 ml / 2 fl oz double cream, 40 g / 1 oz butter and 1 teaspoon of lemon juice. Heat until the sauce is creamy, and stir through 15 g / 0.5 oz sorrel leaves. Grill the salmon for a few minutes on each side, depending on thickness. Serve with the sauce and garnish with wood sorrel flowers.

In this little plant, there is a delicacy of structure superior to what we observe in most

W. Curtis

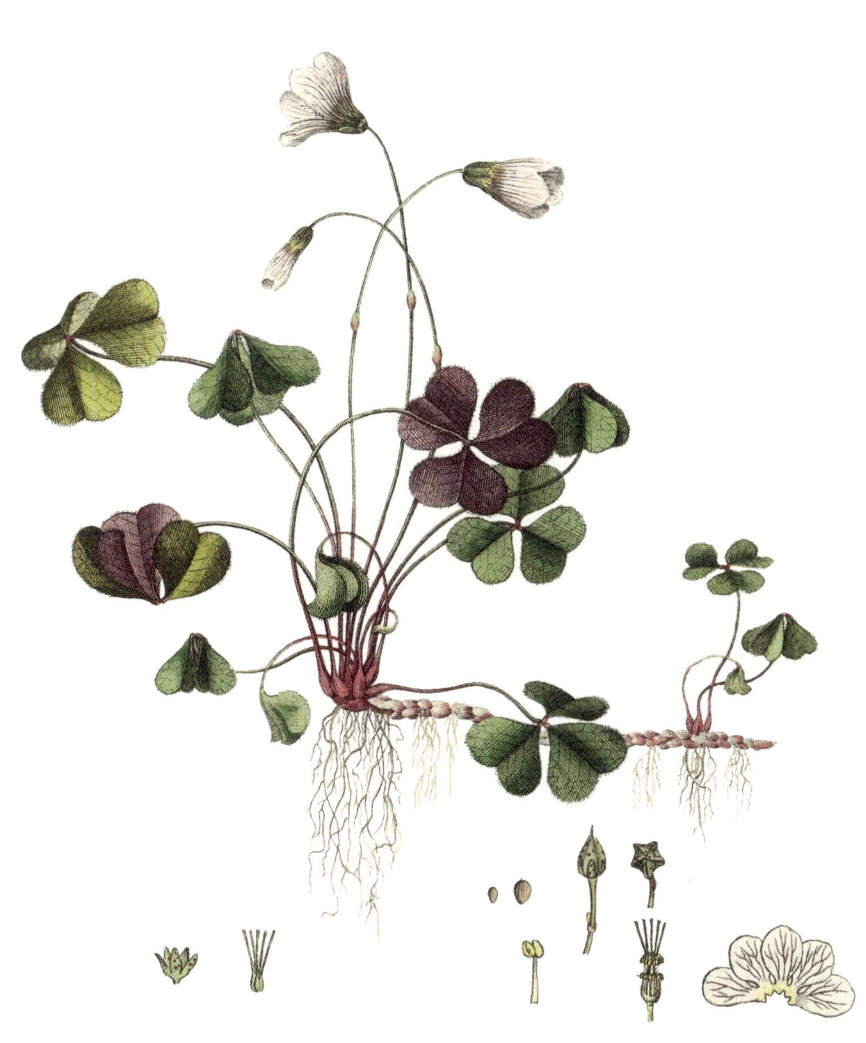

Oxalis acetosella

LADY'S BEDSTRAW

Galium verum

Creeping camouflaged through grass, it's only in summer that lady's bedstraw launches its assault on the senses. It can turn an abandoned carpark into a blinding battery of yellow flowers which thicken the air with their intense honey scent. If the barrage is enough to make you want to lie down, then you're in luck. When dried, the plant was used to stuff mattresses (the clue is in the name). Not only comfortable, its pleasing smell of hay also guarded against fleas. Unlikely as it seems, its talents extend to cheesemaking, as a vegetable rennet and dye. This is a hardworking lady, indeed.

ANATOMY NOTES

STEM — The tradition of placing lady's bedstraw into your shoes is supposed to stop you getting blisters – and to fend off demons too.

FLOWERS — The tightly packed sprays of yellow flowers were used to curdle milk to make cheese in Cheshire. Double Gloucester was coloured its characteristic mild orange using its dye. In recognition of the plant's cheesiness, its scientific name *Galium* is from the Greek word for milk (*gala*).

LEAVES — Make a tea from the leaves to detoxify the liver or use it cool as a facial cleanser. The crushed leaves can be applied to a wound to encourage healing.

SEEDS — Lady's bedstraw is in the same plant family as coffee, and its seeds make a good substitute for the beans when they are roasted, although sadly none of the caffeine.

ROOTS — Chopped up roots produce a stunning red clothing dye when simmered. In the Hebrides it's used to dye wool.

BOTANIST'S RECIPE

CREAM CHEESE — Bruise lady's bedstraw flowers and stems, and cover with water in a pan. Simmer for 35 minutes and strain, retaining the liquid. Add 150 ml / 5 fl oz of the liquid to 1 litre / 34 fl oz of warm whole milk, gently stirring now and then until the milk curdles. You may need to leave it somewhere warm overnight. Scoop the curds into a muslin bag to drain, adding a pinch of salt to help draw out moisture. Remove the muslin and store in the fridge.

ESSENTIALS

Aka — Yellow bedstraw, cheese rennet

Family — Rubiaceae (coffee)

Urban habitat — Grassy patches and verges, meadows, chalky and sandy soils

Height — 50 cm / 1 ft 8 in

Flowers — June to August

Uses — Cheesemaking, mattress stuffing, dye, coffee substitute, medicinal, cleanser

Grow — Sow seed in autumn

Harvest — Flowers in summer and seeds in autumn

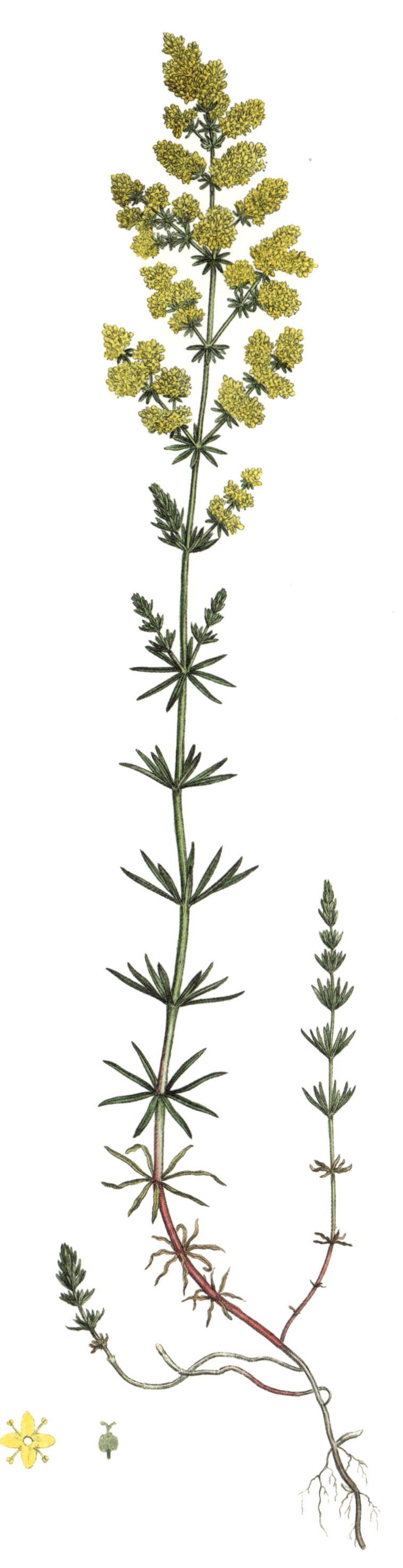

Galium verum

NETTLE

Urtica dioica

No other plant has made so many children cry so hard for so long. But from the nettle's point of view, it needs to bristle with stings filled with burning acids – otherwise such a highly nutritious and useful plant would be eaten, picked, grazed and harvested out of existence. It's an effective deterrent for this true all-rounder: nettle recipes abound for foragers, from tapas to tarte Tatin; you can get dyes, textiles and papers from it; some even use the stings therapeutically to stimulate circulation. And nettles are hugely important for butterflies and ladybirds too, so those childhood tears aren't shed in vain.

ANATOMY NOTES

SEEDS — Birds eat the dried seeds in autumn, but you can gather a few too and add them to your homemade bread for an energy boost.

STEM — Like the leaves, the stems are covered in tiny hollow hairs called trichomes. When touched, they break off, releasing acids and histamine into the skin which we feel as a sting. Roman soldiers rubbed the stings all over their bodies to keep warm in winter. The fibrous outer parts of the stems can be pulped and made into paper.

LEAVES — Wear gloves to harvest nettle leaves, and go for the youngest four to six at the top of the stem. Keep the gloves on to wash them, but after blanching on boiling water the stings are destroyed. Nettle tea has anti-inflammatory properties, good for hay fever sufferers. The dark green dye produced from the leaves was used for camouflage during the Second World War. Butterfly larvae eat the leaves, being small enough to nibble between the stings, so if you're foraging always leave plenty for the wildlife.

BOTANIST'S RECIPE

VEGAN RISOTTO — You can use the first of the nettles and fresh peas for a cheerful early springtime risotto. Sauté a diced red onion in oil and add 200 g / 7 oz of risotto rice. Stir to coat and slosh in a glass or two of vegan white wine. Pour 500 ml / 17 fl oz (or more) of stock bit by bit into the pan, topping up only as the mixture dries out, stirring regularly to stop the rice sticking. When cooked, add 200 g / 7 oz chopped nettle tops and 200 g / 7 oz peas, plus 8 tablespoons of nutritional yeast and cook for 3 minutes. Season, and serve with fresh parsley.

ESSENTIALS

Aka — Stinging nettle

Family — Urticaceae (nettle)

Urban habitat — Disturbed ground, and pretty much everywhere

Height — 1.5 m / 4 ft 11 in

Flowers — May to October

Uses — Food, textile making

Grow — No need to sow seed as it's so common

Harvest — Young leaves in early spring, seeds in autumn

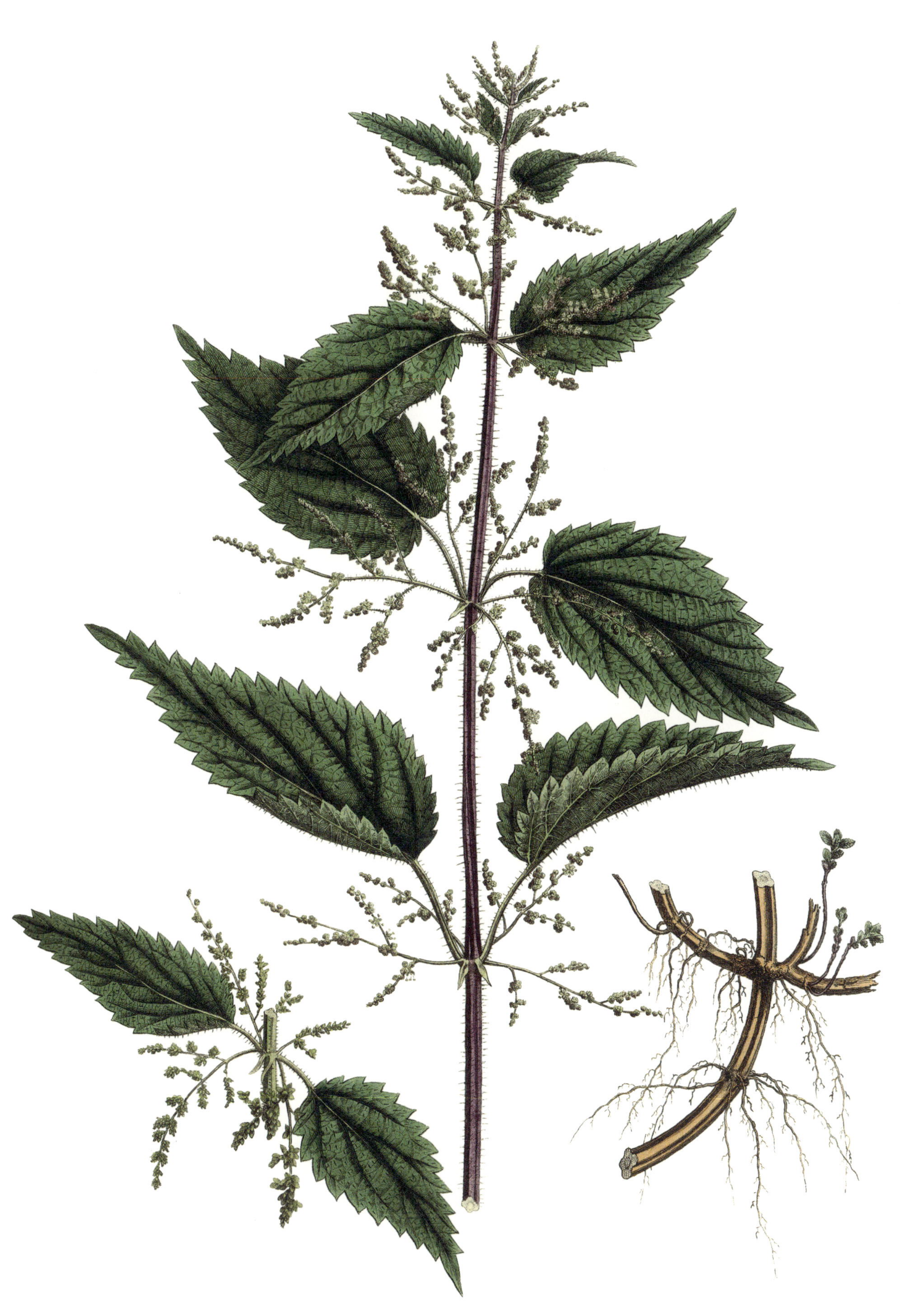

Urtica dioica

MARSH MARIGOLD

Caltha palustris

In ancient custom, the marsh marigold was the unofficial wildflower of May Day – or Beltane as it was called by the Celts – when every doorway was festooned with its sunny flowers in promise of summer to come, and to ward off ill-luck. In some parts of the Britain and Ireland, this practice continues. Less superstitious city dwellers, however, can pickle the flower buds like capers or cook tender leaves to eat like spinach. Do pay marsh marigold respect, though: it can irritate the skin if touched without gloves and it's poisonous raw.

ANATOMY NOTES

FLOWERS — Before the petals completely unfurl, the flowers resemble golden goblets fit for royalty, which is why they're sometimes called kingcups; the scientific name *Caltha* is also derived from the Greek for goblet.

STEM — The sap is toxic and can irritate the eyes, nose and skin. It can blister animals too, and there are reports of it killing cows; the 'marigold' part of its name might actually be a corruption of the Old English *meargalle*, meaning 'horse blister'. Then again, it does look like a marigold and grows in marshes. Roman beggars were said to use it to bring themselves out in a rash of blisters to arouse even more pity.

ROOTS — In some cultures, the root is boiled and mashed and applied to the feet to soothe running sores.

BOTANIST'S RECIPE

GRATIN — This creamy gratin is comfort food at its finest that puts the last of your overwintered parsnips to good use. Boil 450 g / 16 oz (or whatever you can get hold of) of young marsh marigold greens well to break down the toxins. You may need to change the water and repeat until the bitterness is gone. Rinse, press dry, chop and sauté with a sliced red onion and chopped wild thyme. Peel and thinly slice 450 g / 16 oz parsnips, then layer this with the greens in a baking dish. In a pan, heat 350 ml / 12 fl oz of cream and 350 ml / 12 fl oz milk until not quite bubbling, add a pinch of salt and pepper. Pour over the parsnip and greens. Bake in the oven at 190 °C / 375 °F (170 °C / 340 °F fan) for around an hour, or until soft and golden.

ESSENTIALS

Aka — Kingcup

Family — Ranunculaceae (buttercup)

Urban habitat — Damp soils near water

Height — 50 cm / 1 ft 8 in

Flowers — March to May

Uses — Edible flowers, ornamental pond plant

Grow — Buy plants and sink into a pond in spring

Harvest — Buds and young leaves in spring; wear gloves

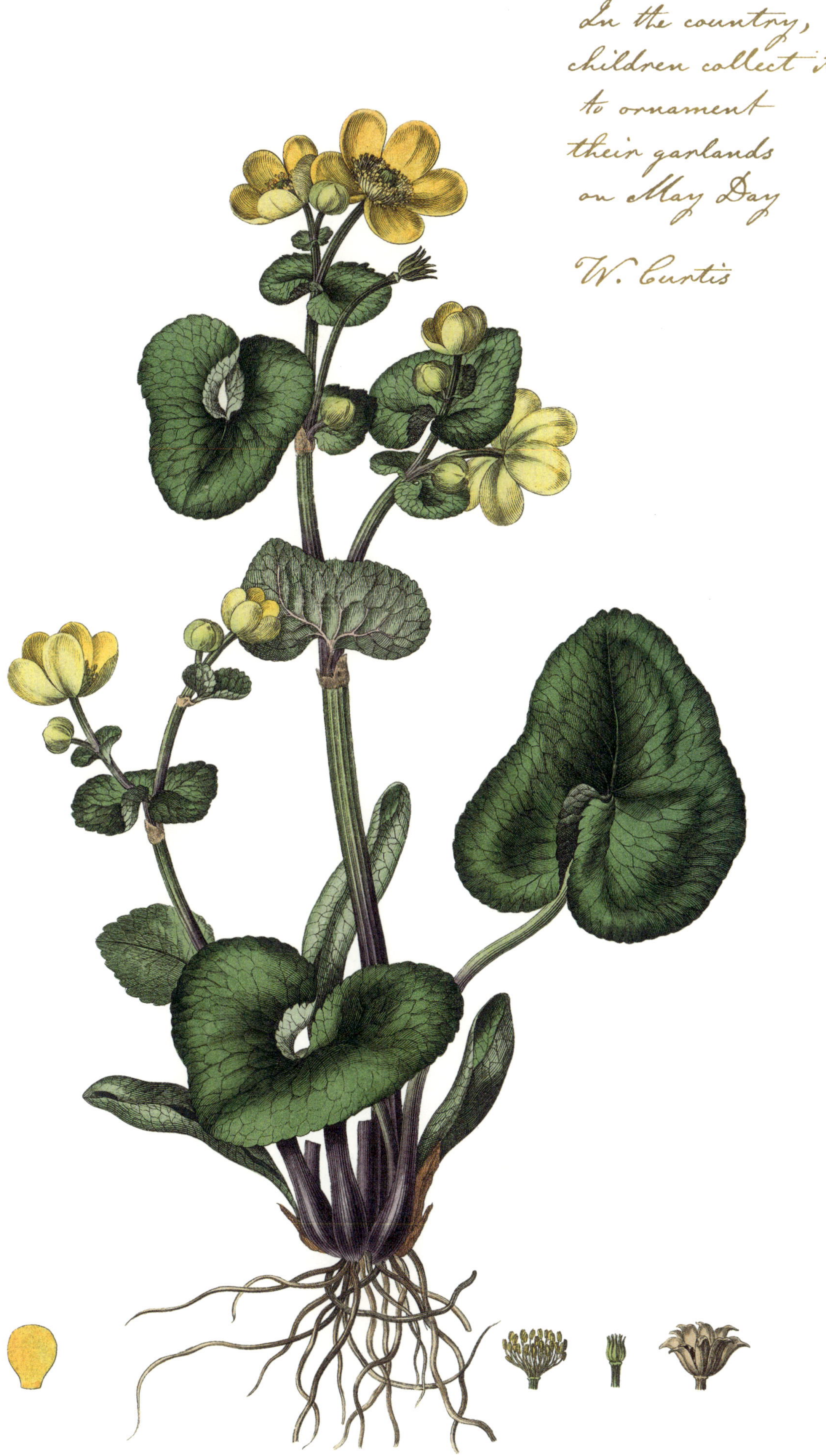

Caltha palustris

WILD CLARY SAGE

Salvia verbenaca

Wild clary sage likes nothing more than basking in the sun, soaking up the heat. In celebration of the height of summer, it sends up aromatic purple blooms in whorls on stiff spikes, which fill the air with an earthy fragrance. You're most likely to see and smell them in graveyards, where they were often grown in medieval times. A versatile herb like its cousin sage, the leaves and flowers are edible and can be used fresh or dried in savoury and sweet dishes alike. Its leaf tea promotes calm and sleep, aids digestion and has anti-inflammatory properties. A beautiful and beneficial plant.

ANATOMY NOTES

FLOWERS — The deep-bellied flowers are a magnet for bees and other pollinators – although wild clary sage can also pollinate itself without visitors if necessary.

SEEDS — The herbalist, Nicholas Culpeper, advised soaking the seeds in water to make a thick solution useful for getting dust out of your eyes – presumably, provided you can still see enough to make it.

LEAVES — An infusion of the leaves was said to strengthen eyesight.

BOTANIST'S RECIPE

FRITTERS — Make the batter several hours before cooking. Mix 100 g / 4 oz flour with a pinch of salt, 2 tablespoons of oil and the zest of a lemon. Whisk in enough warm water to give a consistency of custard. Just before cooking, fold one beaten egg white through the batter. Wash a good handful of wild clary sage flowers and lemon balm leaves. Dry and dip into the batter. Deep fry, drain off excess oil, dip in sugar and serve with lemon sorbet.

ESSENTIALS

Aka — Wild clary, wild sage

Family — Lamiaceae (mint)

Urban habitat — Dry, sunny sites

Height — 60 cm / 2 ft

Flowers — June to September

Uses — Edible flowers and leaves

Grow — Sow seed in autumn

Harvest — Harvest in summer for drying

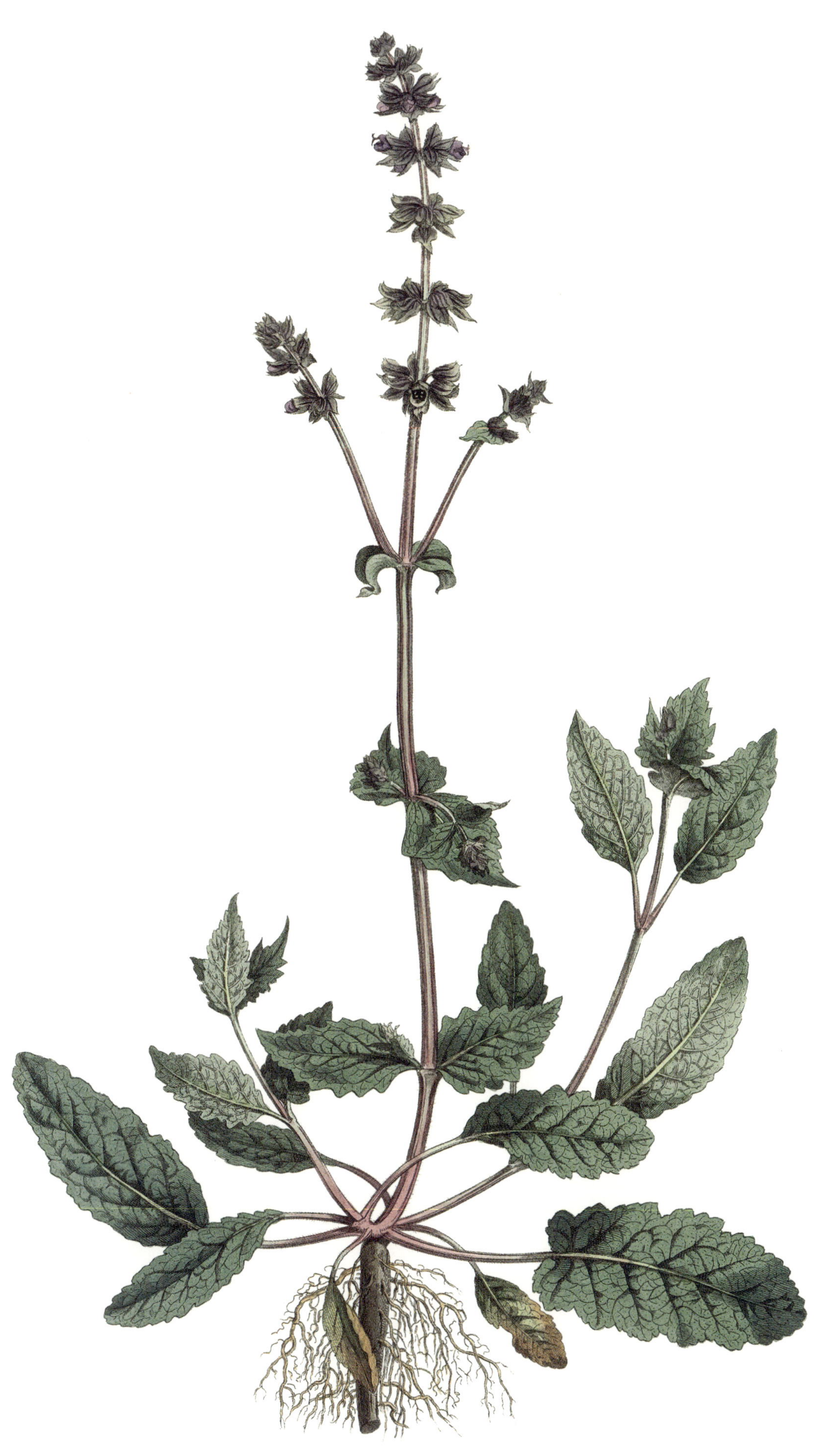

Salvia verbenaca

TREE OYSTER MUSHROOM

Pleurotus ostreatus

Growing on the sides of trees in closely layered steps, smooth and rounded like polished stone, this mushroom is decidedly sculptural. For a fungus, it's a beauty to behold – but it's even more beautiful to eat. You can find them throughout the year, but get them young before the cap completely flattens out, when their substantial meaty texture and subtle nutty flavour, touched with a hint of anise, is at its best. High in protein and vitamin C, they are now grown commercially worldwide, taking a particularly prominent role in Oriental cuisine.

ANATOMY NOTES

CAP — The mushroom contains a statin used to treat high cholesterol. Research is also underway into possible uses against HIV and sarcoma cancers.

MYCELIUM — Caution, this mushroom is carnivorous! Its mycelium which acts a little like roots, searching out nutrients and water, consumes tiny nematode worms, as well as helping to decompose decaying wood. In contaminated areas it can break down pollutants, helping to clean the soil (a process called mycoremediation), but the impurities can reside in the cap, making it unsafe to eat. Amazingly, the tree oyster's mycelium can also be turned into tough, leathery textiles with a range of uses, from packaging to clothing.

BOTANIST'S RECIPE

KEBAB — Rustle up this spicy vegetarian alternative to the classic doner kebab by dry-frying 500 g / 18 oz sliced tree oyster mushrooms for two minutes. Now add some chilli or garlic oil, 2 finely chopped garlic cloves, 1 teaspoon of cayenne pepper, 1 teaspoon of smoked paprika, 1 teaspoon of ground coriander, 1 teaspoon of cumin salt, some twists of pepper, and stir to coat the mushrooms. Add a little water to help cook the mushrooms, stirring for a minute or so. Serve with warm pitta bread with shredded cabbage, chilli sauce, chopped tomatoes and minted yoghurt.

ESSENTIALS

Aka — Pearl oyster mushroom, hiratake

Family — Pleurotaceae

Urban habitat — Deciduous trees

Width — 20 cm / 8 in

Uses — Food

Grow — Try growing from spores in autumn

Harvest — All year round, before the cap flattens

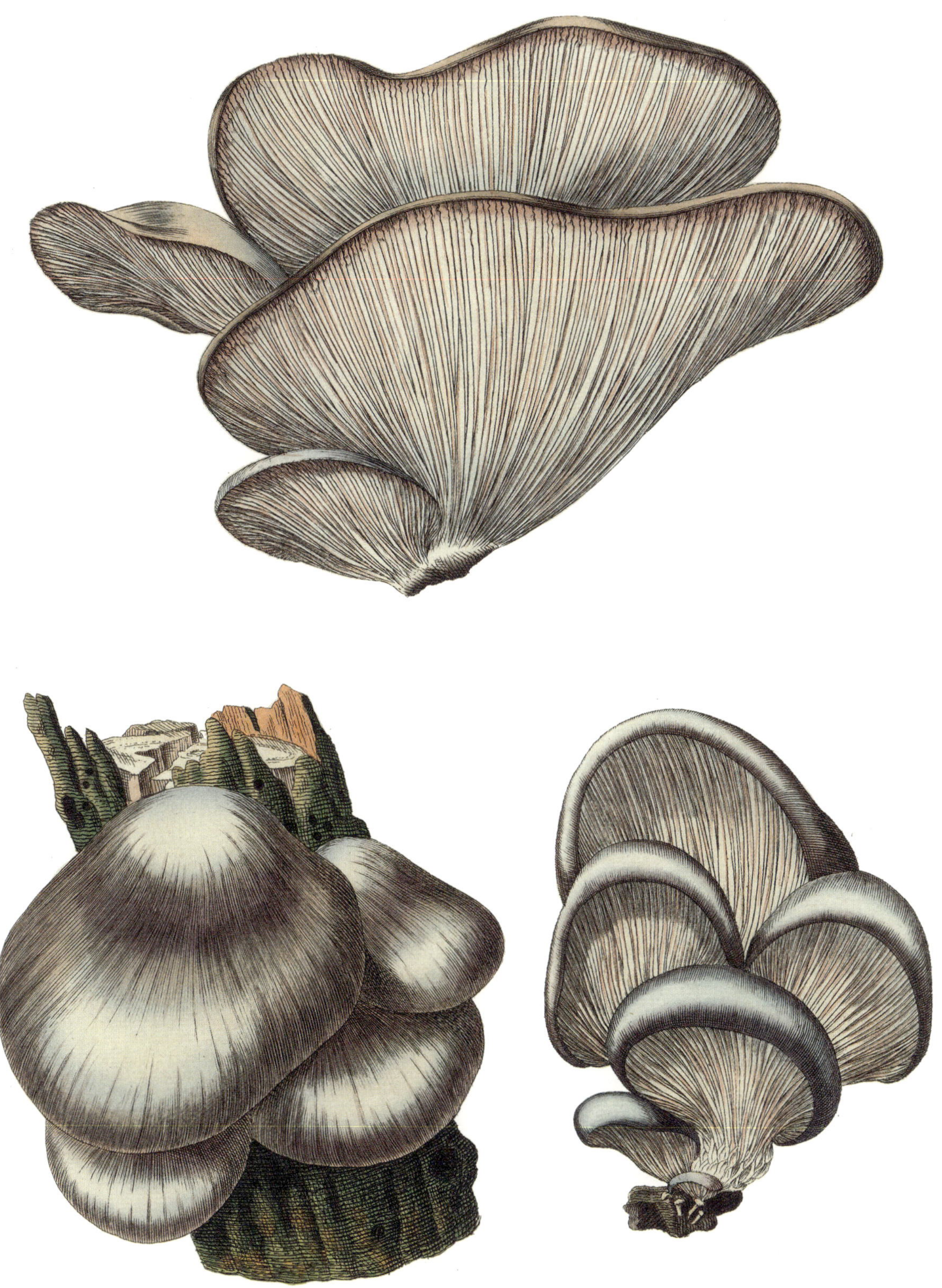

Pleurotus ostreatus

BITTER VETCH

Lathyrus linifolius

In spring, the delicate pink flowers of bitter vetch quiver by the roadside with every passing car. Beneath the soil, however, the plant hides deliciously sweet and nutty edible tubers. Before the introduction of the potato, they were eaten in the Scottish Highlands, dried out (to remove the plant's mild toxins) and chewed like liquorice or tobacco, often with a tipple of whisky. Bitter vetch was particularly valued when food was scarce for its hunger-suppressing properties – something that's arousing interest again now in the never-ending quest for the perfect slimming aid.

ANATOMY NOTES

SEEDS — Once cooked to destroy its natural toxins, the seeds can be eaten and taste a little like sweet chestnuts. The only problem is that they're tiny and not worth the backbreaking effort of gathering enough for a decent meal.

TUBERS — The Scots called the dried tubers 'cairmeal', and fermented them as a kind of medicinal beer, especially used to treat throat problems. As rumour has it, King Charles II fed it to his mistresses to keep them in shape.

ROOTS — As with most plants in the pea family, the roots of the bitter vetch 'fix' nitrogen from the air, converting it into a nutrient in the soil. It's therefore a great choice for improving soil health.

BOTANIST'S RECIPE

ROAST TUBERS — Fill up on these with your next Sunday roast and you won't need any pudding. Peel tubers and place in a pan of salted water. Bring to the boil and simmer hard for a few minutes. Meanwhile, fill a roasting tin with about 0.5 cm / 0.2 in of oil and place into an oven pre-heated to 200 °C / 390 °F (180 °C / 355 °F fan) for 5 minutes. Dry the tubers, place carefully into the hot tray with cloves of garlic, coating with the oil, and season with salt and pepper. Roast for 40 minutes, turning once or twice, until golden and crispy.

ESSENTIALS

Aka — Heath pea

Family — Fabaceae (pea)

Urban habitat — Roadsides, verges and woods

Height — 30 cm / 1 ft

Flowers — May to July

Uses — Edible root once cooked

Grow — Plant tubers in spring

Harvest — Tubers in autumn

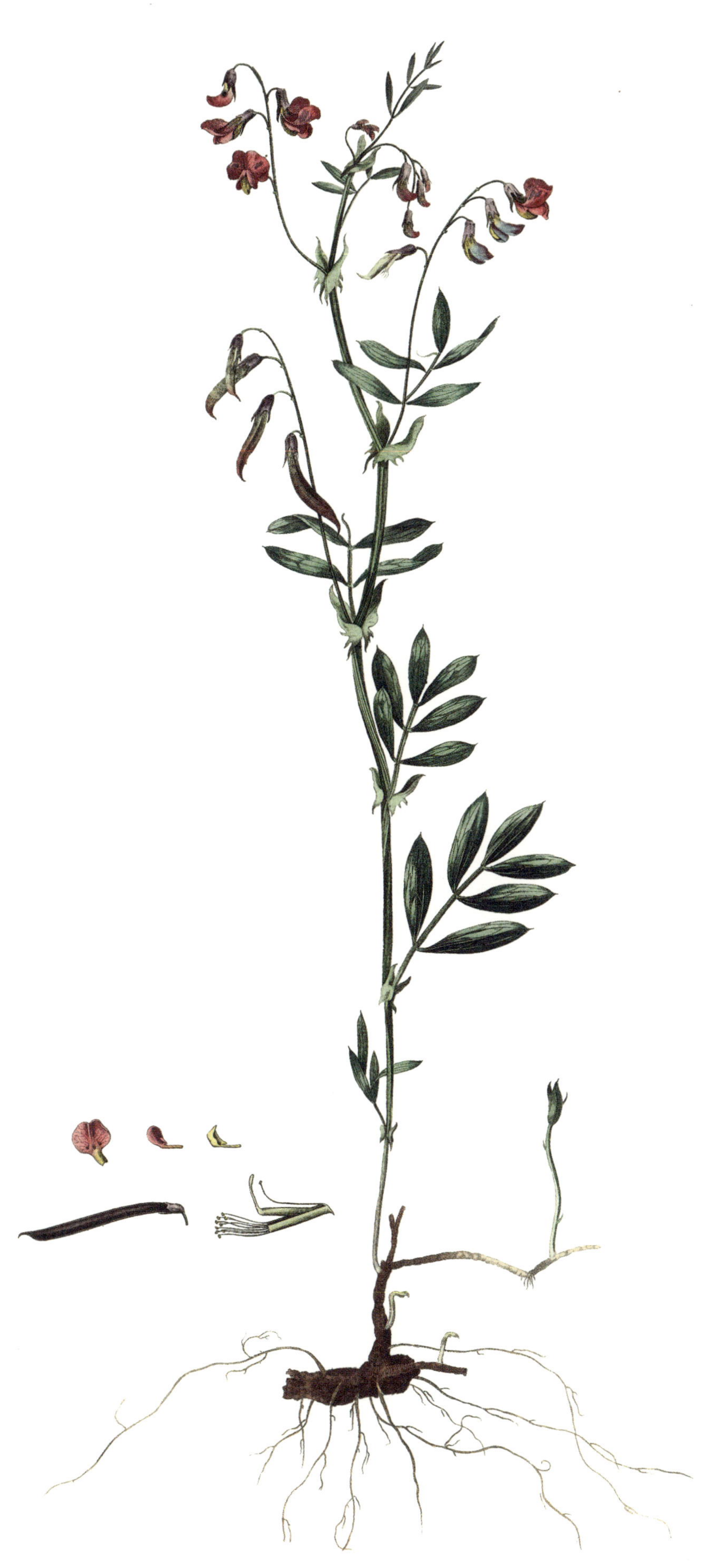

Lathyrus linifolius

JACK-IN-THE-HEDGE

Alliaria petiolata

Many people would like garlic a lot more if it wasn't so overpoweringly, well, garlicky. Step forward jack-in-the-hedge, the edible plant with only a subtle garlic flavour. Appearing beneath hedges – or in broken concrete by fences in town – it features dancing white flowers and heart-shaped leaves that give off a delicate aroma of garlic when caressed. Chop them and add raw to salads, salsas, soups, sauces and dips for that mild garlic taste flamed by a little mustard heat – spice that becomes overwhelming if you pick them after the plant blooms.

ANATOMY NOTES

SEEDPOD — The seeds develop in long, green seedpods and can be used in the kitchen as a substitute for mustard seeds. Birds love to snack on them too.

FLOWERS — Its pretty cross-shaped flowers beautifully set off a salad, but alliumphobes (yes, those fearful of garlic) be warned: they're very garlicky.

LEAVES — The nettle-like leaves towards the top of the plant are the most tender and are best harvested in the first year of its biennial life cycle. They usually taste better raw than cooked, especially when used shortly after harvesting. High in vitamins A and C, they're said to strengthen the digestive system.

STEM — The plant usually throws up only one shoot, but if it's damaged or cut, it may put out many other shoots in reaction.

ROOTS — You can use the roots to make horseradish sauce, first making sure to discard the fibrous woody parts. Their 'S' shape helps them to cling to slopes and steep verges.

ESSENTIALS

Aka — Garlic mustard

Family — Brassicaceae (cabbage)

Urban habitat — Underneath hedges

Height — 70 cm / 2 ft 4 in

Flowers — April to June

Uses — Food

Grow — Sow seed in spring

Harvest — Early spring

BOTANIST'S RECIPE

DOLMADES — Instead of using vine leaves, try jack-in-the-hedge for a left-field, but mouthwatering, springtime appetiser. Boil the leaves for 4 minutes, then drain and squeeze the water out. Make a filling of your choice, such as cooked rice with chopped sundried tomatoes and olives, pine nuts and basil. Place the filling into a leaf and roll up. Serve hot or cold.

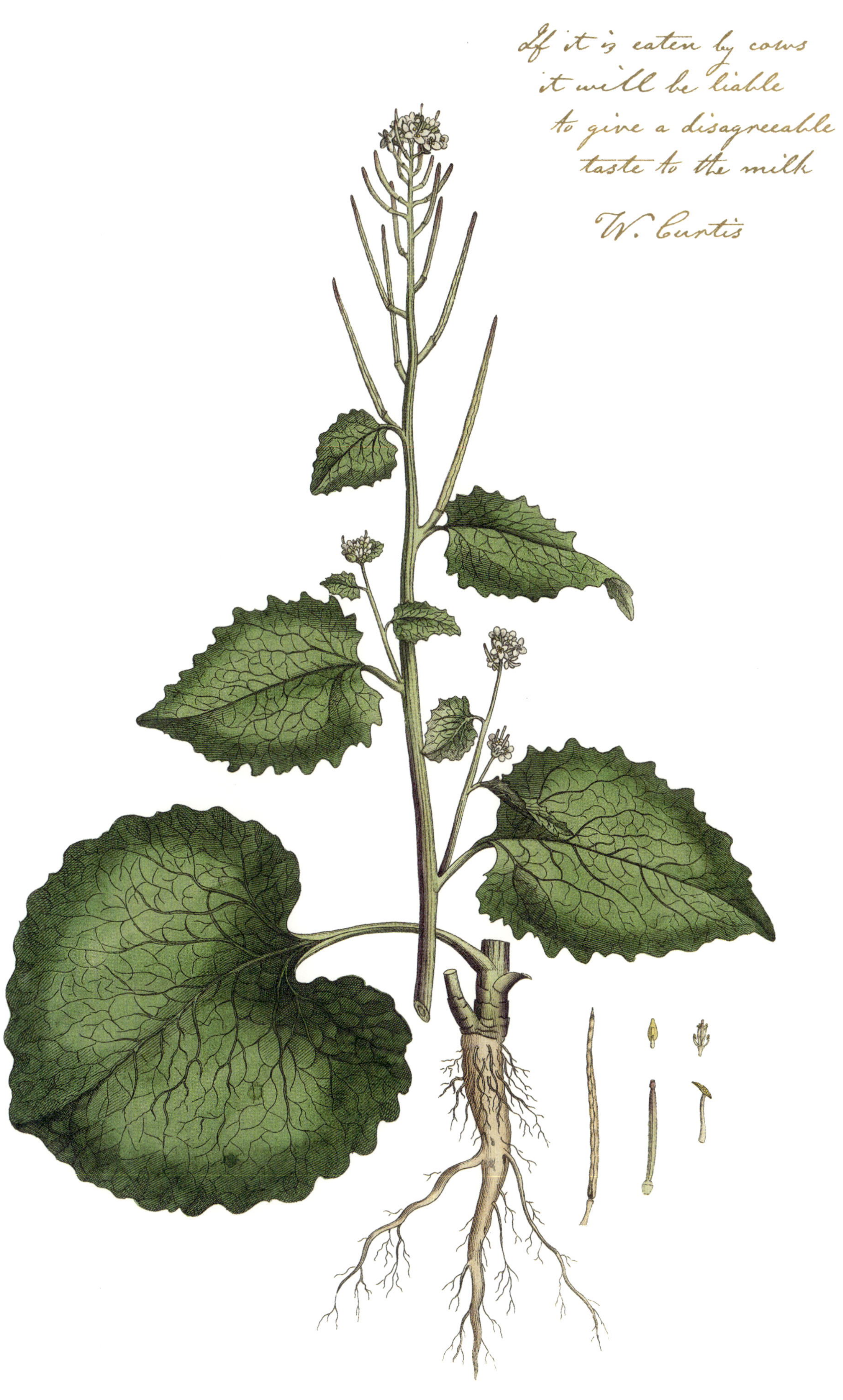

Alliaria petiolata

SOW THISTLE

Sonchus oleraceus

If you thought dandelions were everywhere, their rangy big brother, the sow thistle, really does get around. In even the tiniest tuffets it can make itself at home beneath a road sign at the noisiest junction in the busiest city. To many the archetypal weed, it defends itself with prickly leaves – which are surprisingly appetising if eaten young. They taste like a sweet lettuce and are loaded with nearly twice the vitamin C of spinach. Peel the stems and you have ersatz asparagus. But do mind the location of your crop – sow thistles like growing at the bottom of lampposts as much as dogs enjoy watering them.

ANATOMY NOTES

FLOWERS — Young buds and flowers can be added to salads or battered into fritters.

STEM — The Greek word *sonchus* means hollow and refers to the stem. Pigs and hares love to eat this plant, hence its common names, and farmers thought it was part of a healthy diet for lactating sows.

LEAVES — When harvesting, avoid prickly leaves by using younger ones higher up the stems. Trim any prickles off before eating, raw or cooked. The sap from the leaves is used by Māoris to make a bitter chewing gum; in fact, the plant, called *puha*, is used widely in Māori cuisine and culture, including the popular dish 'pork and *puha*' – pork bones stewed with sow thistle, sweet potato and other vegetables.

ROOTS — The roots can be roasted and eaten, although they are quite small for all the effort.

BOTANIST'S RECIPE

STIR FRY — Fresh spring sow thistle leaves go very well with pork, appropriately enough. Make a marinade of soy sauce, water, white wine, pinches each of sugar and corn flour. Slice around 300 g / 11 oz pork tenderloin, season, and place in the marinade for around an hour. Heat a dash of oil in a large pan and stir fry the pork for a couple of minutes to seal. Remove from the pan and add the veg of your choice in order according to cooking times, finishing with chopped garlic, spring onions and chopped sow thistle greens. Then return the pork to the pan, add the marinade and cook through for another minute or two. Serve with rice.

ESSENTIALS

Aka — Hare's lettuce

Family — Asteraceae (daisy)

Urban habitat — Verges, roadsides, disturbed and waste ground

Height — 70 cm / 2 ft 4 in

Flowers — July to September

Uses — Food

Grow — Sow seed; but it's likely to be growing in your garden already

Harvest — Young leaves in spring

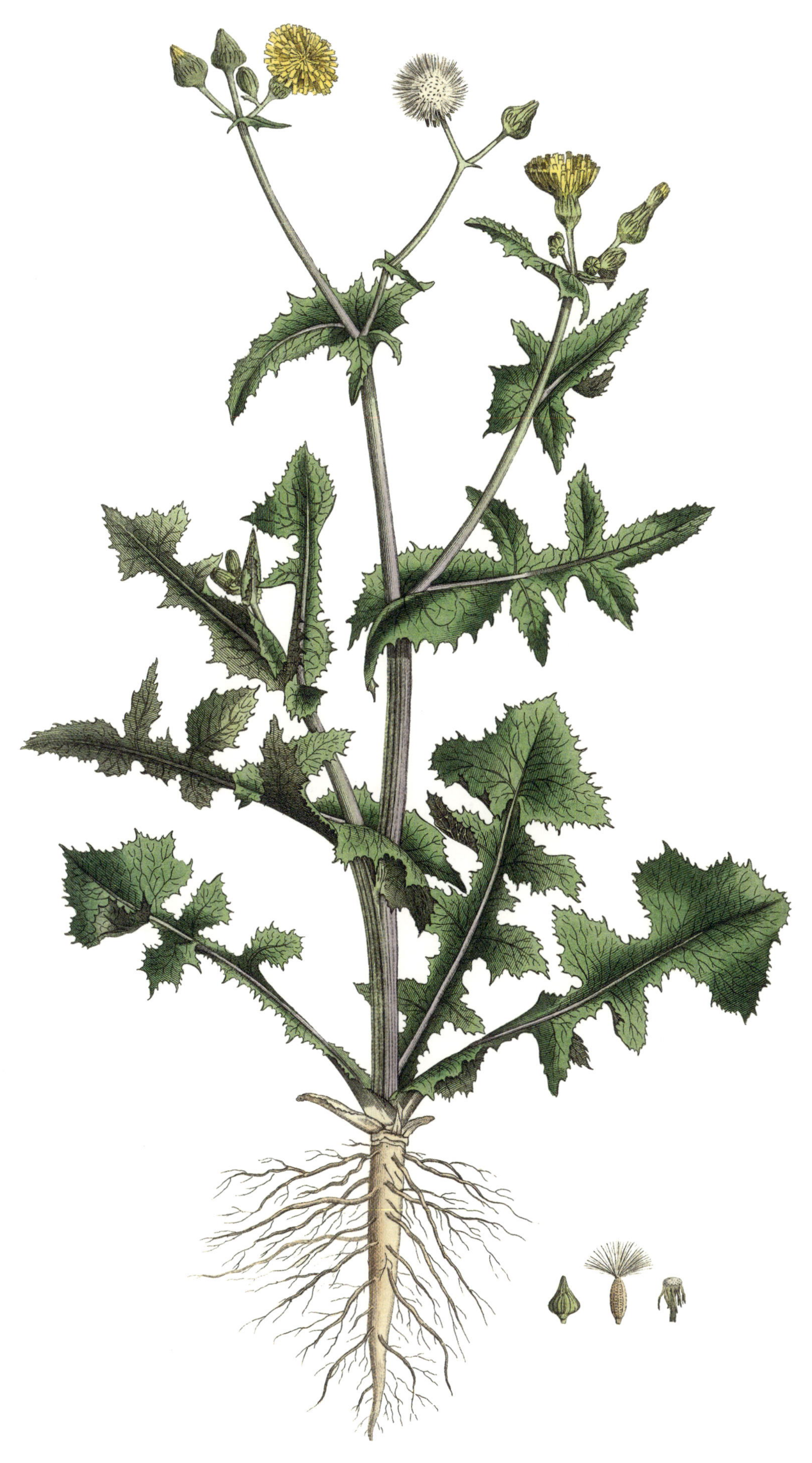

Sonchus oleraceus

CORN POPPY

Papaver rhoeas

Once largely confined to cornfields, this striking red poppy is now quite at home in the city. You'll see them standing to attention along city roadsides like soldiers in scarlet tunics – about right for our most famous floral symbol of remembrance. In autumn, they sprout urn-shaped casings filled with seeds, which you can use to decorate and flavour bread, cakes and pastries. Young leaves add dimension to salads, which could themselves be dressed with poppyseed oil. So, it's not just a flower of enormous cultural significance; it's pretty handy in the kitchen too.

ANATOMY NOTES

LEAVES — If you're gathering tender young leaves to eat, make sure you're doing it before the flower buds form, when the leaves become toxic.

FLOWERS — Long before its associations with the First World War, the corn poppy was a symbol of agricultural fertility. The Romans depicted poppies and corn in images of Ceres, goddess of agriculture. The vibrant crimson colouring extracted from the flowers is used by vintners to deepen the hue of red wines. Flower infusions have been used to soothe mild pains like toothache, as a sedative, or taken for irritable coughs.

SEEDS — A single plant produces 60,000 seeds, and each seed can lie dormant in the soil for 80 years, waiting to spring to life when disturbed. Following the carnage of war, the poppy is one of the first plants to grow on battlefields, making a symbolic sea of blood-red flowers.

BOTANIST'S RECIPE

MOHNKUCHEN — Mix 170 g / 6 oz plain flour, 70 g / 3 oz sugar and a pinch of salt. Rub in 80 g / 3 oz cold butter and press the mixture into a lined cake tin; chill. Dissolve 100 g / 4 oz sugar in 500 ml / 17 fl oz milk and 100 g / 4 oz butter. Bring to the boil and gradually add 100 g / 4 oz semolina and 120 g / 4 oz poppy seeds, stirring constantly until thick; then cool. Beat 250 g / 9 oz mascarpone, 1 teaspoon vanilla extract and 1 egg until smooth, combine with the poppyseed mixture, and pour over the base. Mix 50 g / 2 oz flour, 50 g / 2 oz butter, 30 g / 1 oz soft brown sugar into a crumble to top off. Bake at 180 °C / 355 °F (160 °C / 320 °F fan) for 40 minutes until cooked and golden.

ESSENTIALS

Aka — African rose

Family — Papaveraceae (poppy)

Urban habitat — Pavement cracks, roadsides, grassy margins

Height — 70 cm / 2 ft 4 in

Flowers — June to August

Uses — Edible seeds, analgesic, sedative, food dye, cultural symbol

Grow — Sow in autumn

Harvest — Collect seed in autumn, leaves in spring

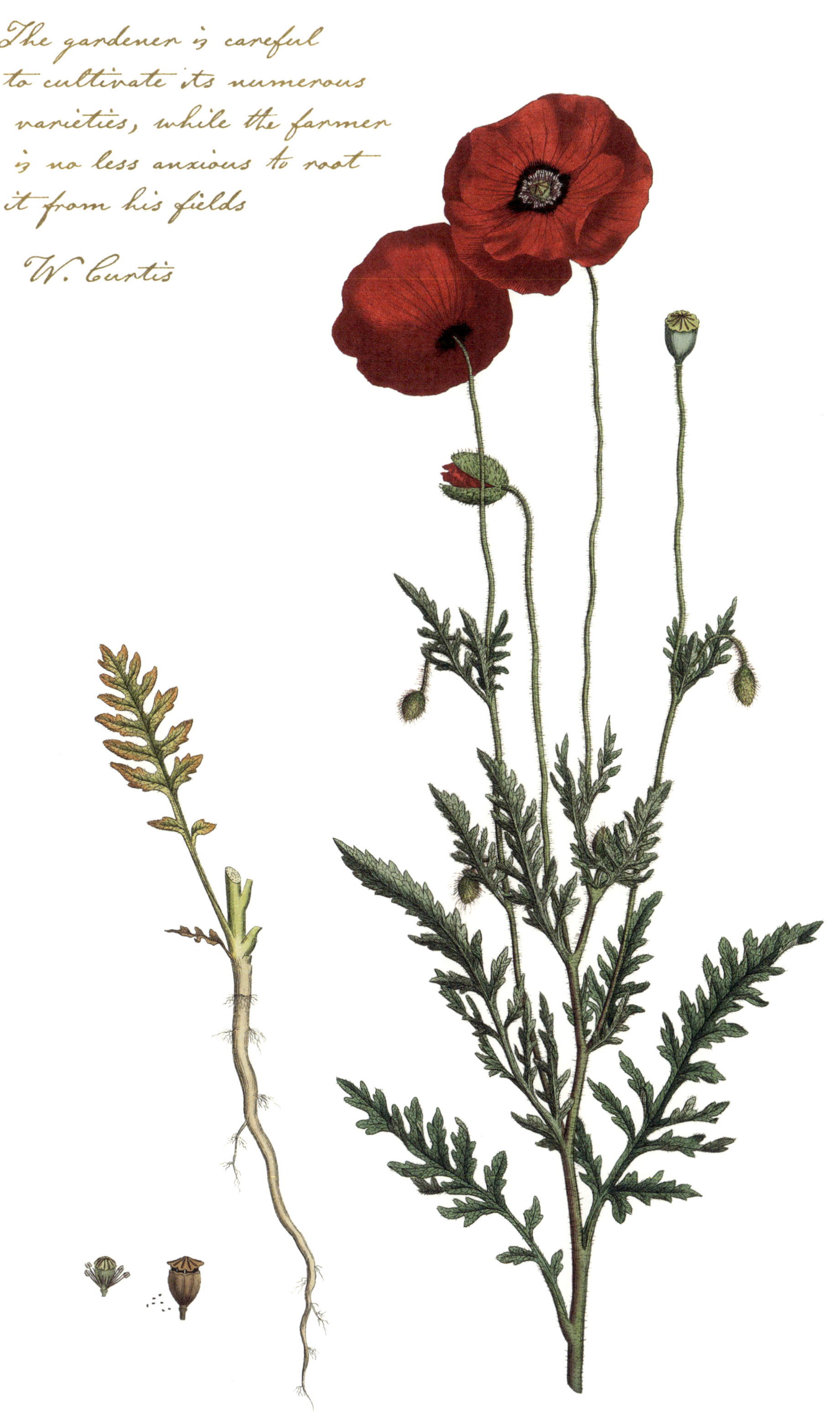

Papaver rhoeas

O R E G A N O

Origanum vulgare

Have you ever walked by a patch of scrubby waste ground and caught a scent that suddenly transports you to a Greek island? That smell, so evocative of Mediterranean summer heat, is oregano – an aromatic plant that is just at home growing in cooler climates. The wild version is often more flavoursome than shop-bought stuff, which is sometimes mixed with other *Origanum* species. Its leaves, stem and pretty pink flowers can be used fresh or dried, in drinks, cooking and baking. In the garden it releases its scent every time you brush past, attracts bees and other pollinators – and reminds you that it's high time for another pizza.

A N A T O M Y N O T E S

F L O W E R S — Oregano tea made from the flowers and leaves can help you fight off colds and sore throats. Before hops became popular, oregano flowers were used to flavour beer; their bitterness masked any unpleasant tastes effectively. It was also supposed to get you drunk more quickly.

S T E M — The stem produces a deep brown to black natural dye. In Ancient Greece, oregano growing on a grave was taken as a sign that the departed was content in the afterlife.

B O T A N I S T ' S R E C I P E

C A K E — A lemon drizzle cake with hidden depths, courtesy of an oregano hit that really brings out the sweet citrus. Cream 250 g / 9 oz of soft butter with 250 g / 9 oz caster sugar. Mix in 4 large beaten eggs, a little at time. Mix in 250 g / 9 oz of sifted self-raising flour and the zest of a lemon, a teaspoon of finely chopped oregano, and a drop or two of vanilla essence. Tip the mixture into a greased cake tin (with greaseproof paper on the base), and bake for around 45 minutes in an oven, preheated to 180 °C / 355 °F (160 °C / 320 °F fan), or until a skewer inserted into it comes out clean. Let it cool, prick with a fork and drizzle over with the juice of the lemon mixed with 70 g / 2 oz of caster sugar. Decorate with oregano flowers.

E S S E N T I A L S

Aka — Wild marjoram

Family — Lamiaceae (mint)

Urban habitat — Dry, sunny verges, grassy patches, scrub and waste ground

Height — 70 cm / 2 ft 4 in

Flowers — July to September

Uses — Edible leaves and flowers

Grow — Sow seeds in spring

Harvest — Leaves spring to autumn, flowers in summer

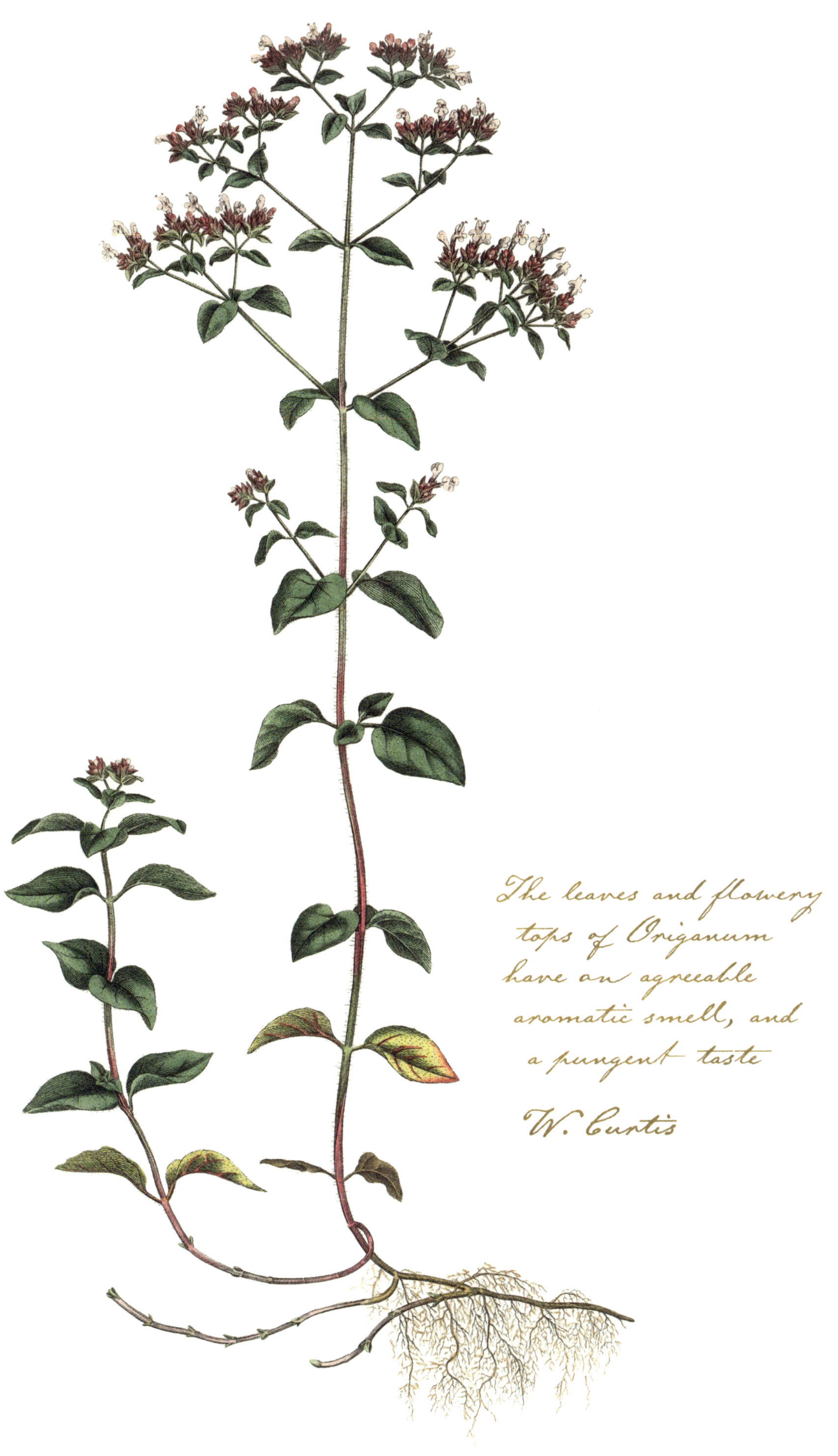

Origanum vulgare

M A L L O W

Malva sylvestris

Mallow is a large and friendly plant, waving its hand-sized leaves at each passing car from roadside verges, while nodding hello with its cheerful pink and purple-veined flowers. It's also a very useful plant, its flowers, seeds, leaves and roots all earning their place in the forager's kitchen. Furthermore, it's mucilaginous, producing a gel-like substance, which can pass for substitute egg (albeit green-coloured), in vegan cooking. And for herbalists, those gels have long been used to make healing ointments and poultices. Seeding freely and growing strongly, it could easily pop up in your back garden – a useful guest.

A N A T O M Y   N O T E S

S E E D S — The puck-shaped seedpods or nutlets look like little cheeses (hence one of its names) and make a quick and convenient snack, packed with healthy fatty acids. A tea of its seeds and leaves soothes inflammation.

L E A V E S — Young spring leaves are edible and high in iron, and great to thicken soups. They're also useful as a poultice for bruises and stings. If you're caught short in the wild, few leaves feel softer.

S T E M — The stem is fibrous enough to be made into cord, paper and textiles.

F L O W E R — Raw flowers add a dash of colour to the salad bowl, and buds can be pickled like capers. Alkalis turn a tincture of the petals from violet to green.

R O O T S — The roots are popular in Chinese cuisine, particularly featuring in soups and broths.

B O T A N I S T ' S   R E C I P E

M A R S H M A L L O W S — Sweetshop marshmallows no longer contain its namesake plant marsh mallow (*Althaea officinalis*) as they once did. Even so, its cousin common mallow also makes a scrummy campfire treat: simmer up husked seeds or torn leaves in twice as much water until it becomes sticky and thick; then stir the strained liquid into an egg already beaten with 35 g / 1 oz sugar. Spoon out on to a tray and bake at 160°C / 320 °F (140°C / 285 °F fan) for 20 minutes.

E S S E N T I A L S

Aka — Cheeses, common mallow

Family — Malvaceae (mallow)

Urban habitat — Grassy roadsides, waste ground

Height — 1.5 m / 4 ft 11 in

Flowers — June to September

Uses — Edible seeds, flowers and leaves

Grow — Sow seed in autumn

Harvest — Leaves in spring, flowers in summer

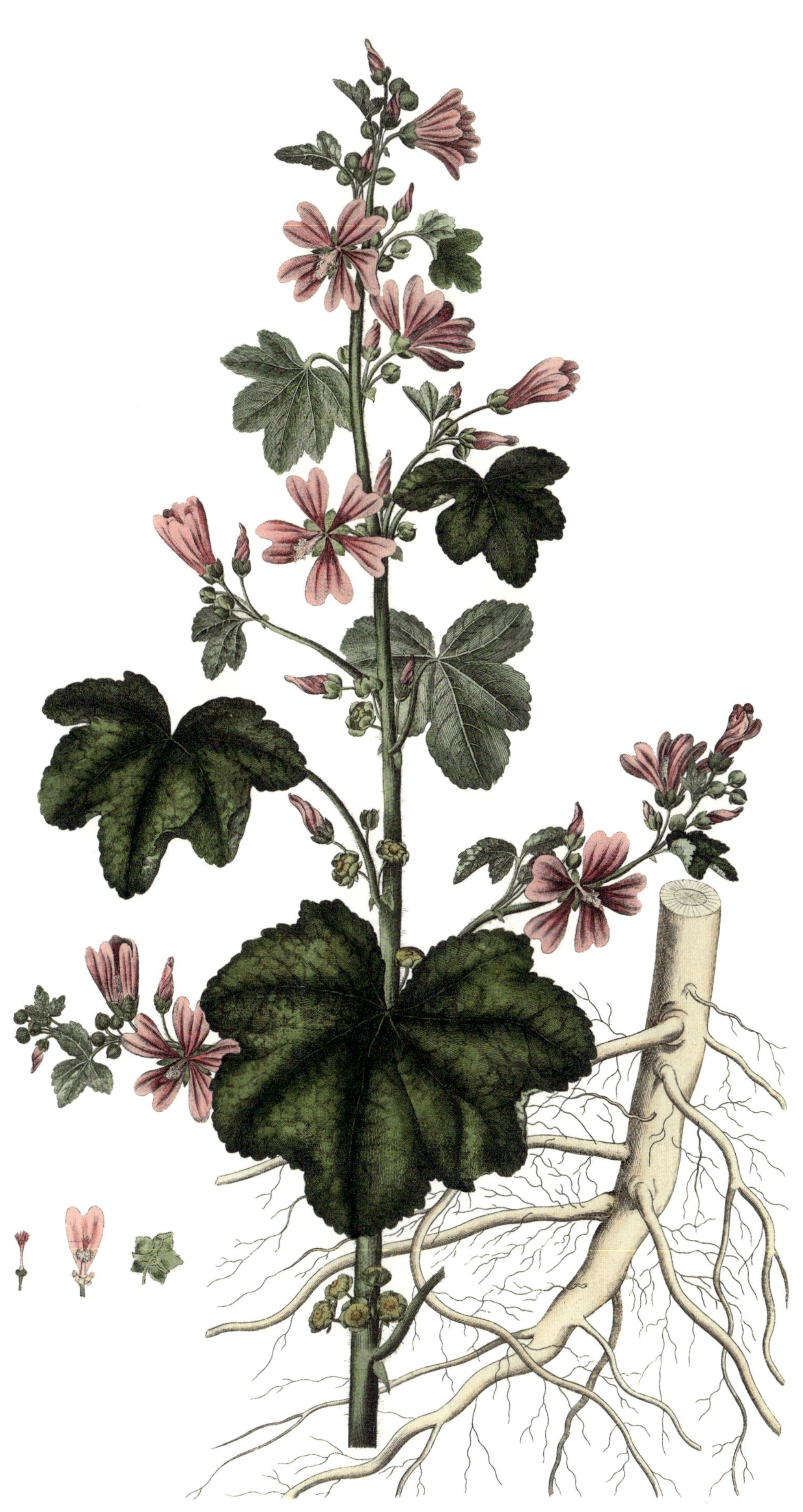

Malva sylvestris

MEADOWSWEET

Filipendula ulmaria

In summer, masses of meadowsweet's creamy flowerheads joyously froth over low walls like champagne bubbles running over a brim – which, for a plant that has long been used to flavour beer, mead, wine and indeed champagne, seems apt. Perhaps it's equally appropriate that it's also a traditional cure for an acid stomach. Meadowsweet's not just for the boozers, though. The gorgeous blooms, smelling of subtle vanilla with almond notes, can be made into fritters or sweet syrup, or simply cut to decorate and perfume the home.

ANATOMY NOTES

LEAVES — Dried leaves can be used as a sugar-free sweetener in tea and drinks.

FLOWERS — A tea of meadowsweet flowers, which contain salicylic acid, an important component of aspirin, acts as a mild painkiller and is prescribed for many ailments including fevers and gout. Queen Elizabeth I wasn't alone in favouring meadowsweet as a strewing herb; scattered on cold floors it fills the home with its lovely scent and keeps your feet warm.

ROOTS — The roots can be peeled and chewed to help ease a headache, and a black dye can be obtained when mixed with copper.

BOTANIST'S RECIPE

CHAMPAGNE — OK, it isn't strictly a champagne, but this fizzy summer cooler is perfect to enjoy chilled as the shadows lengthen on a warm evening. It's gently alcoholic; if you need something a little stronger, it partners gin like a long-lost brother. Simmer a dozen meadowsweet heads in 3 litres / 5.3 pts of water for 15 minutes, then strain the liquid and remove the flowerheads. Bring to the boil and add 400 g / 14 oz of sugar and 12g / 0.5 oz cream of tartar, simmer until dissolved. Pour the contents into a sterile brewing vessel and allow to cool, adding some brewer's or champagne yeast. Leave covered in a cool place for a week while it ferments, then bottle when it's calmed down (you don't want that glass to explode!). Store in a cool, dry place and wait at least another week (several is better) before drinking.

ESSENTIALS

Aka — Meadwort, queen-of-the-meadow

Family — Rosaceae (rose)

Urban habitat — Damp grassy areas

Height — 1 m / 3 ft 3 in

Flowers — June to August

Uses — Flowers for brewing, leaves for sweetening

Grow — Sow seed in spring

Harvest — Flowers in July

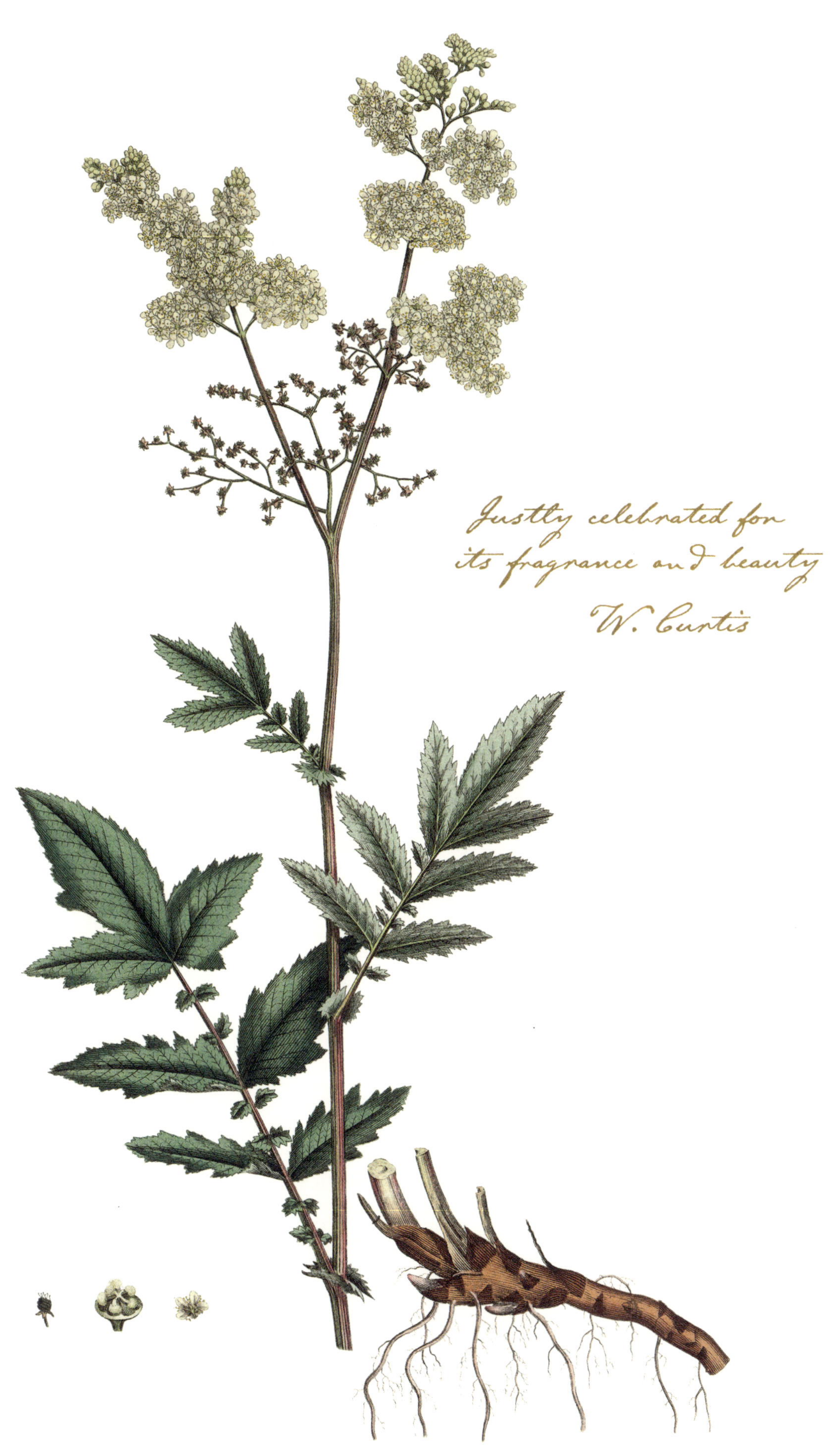

Filipendula ulmaria

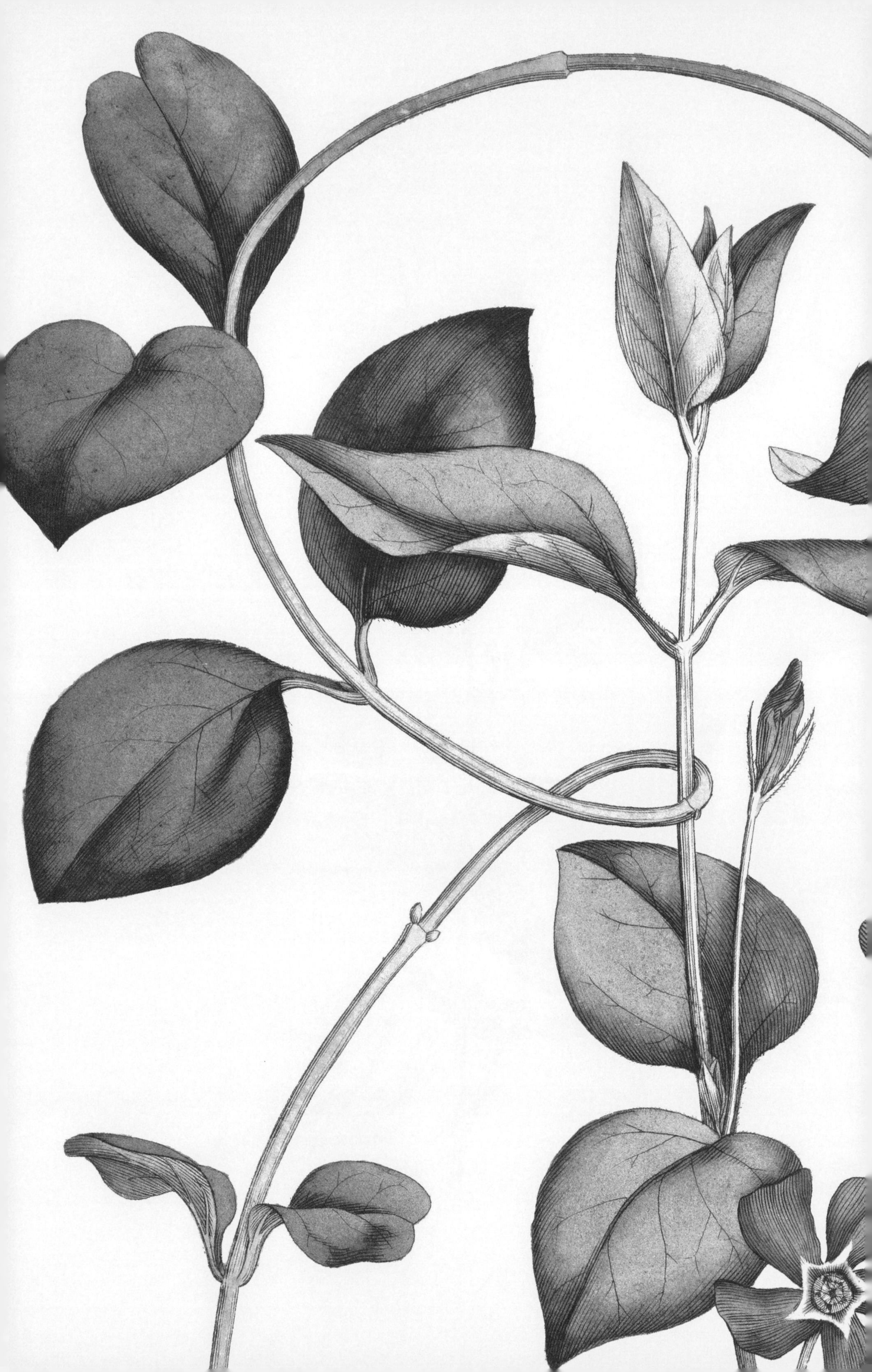

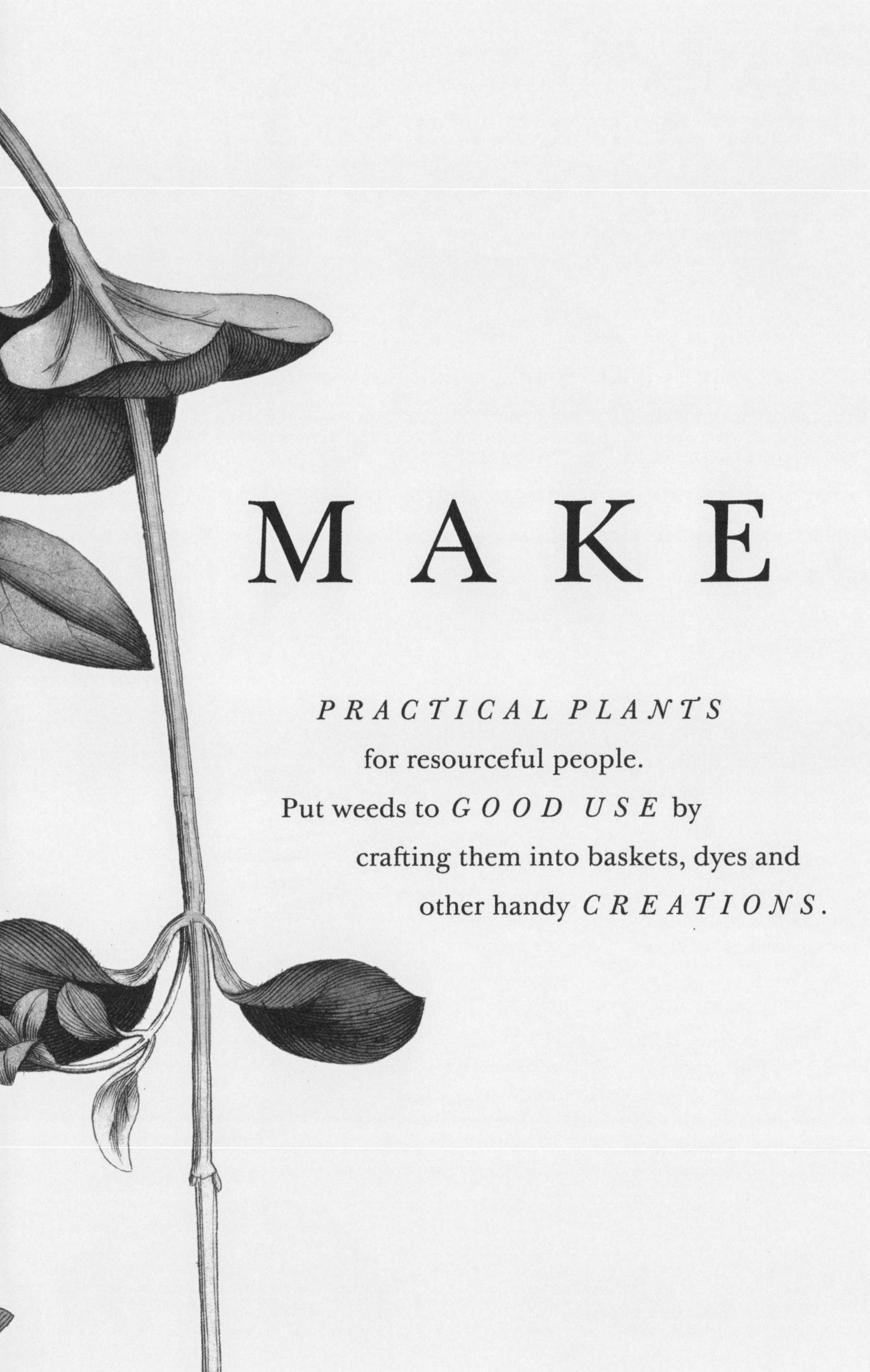

MAKE

PRACTICAL PLANTS
for resourceful people.
Put weeds to *GOOD USE* by
crafting them into baskets, dyes and
other handy *CREATIONS*.

GREATER PERIWINKLE

Vinca major

Blue stars twinkling at you from the shadow of stout trees in your local park is probably greater periwinkle. Sought after for city gardens, it thrives in dry shade and is often used as ground cover for its lush year-round foliage – just stay quick-witted to prevent it from overrunning less vigorous plants nearby. Being poisonous, it's not one you can control by harvesting for the salad bowl, but its long and flexible stems are useful to make into twine or cord for weaving.

ANATOMY NOTES

FLOWERS — For an even greater stellar effect, try the white-flowered 'Alba' cultivar. Whatever the colour, periwinkle flowers will attract pollinators to your garden.

LEAVES — The 'Variegata' version features leaves with light-golden margins. It's a slower grower and easier to control in small garden beds. Old-time herbalists used periwinkle as a laxative, for piles and skin complaints, but not anymore.

STEMS — The stems grow to cover the ground extensively, and put out more roots at their nodes, where the leaves grow, to ensure domination of their competitors.

BOTANIST'S CRAFT

WEAVING — The long stems of periwinkle are perfect for cord, and can be woven into baskets and macramé plant hangers. Gather a handful of the longest stems and strip off the leaves, and you're ready to go. Whatever you're weaving will tighten nicely as the stems dry.

ESSENTIALS

Aka — Grave myrtle

Family — Apocynaceae (dogbane)

Urban habitat — Shady spots under trees and hedges

Height — 20 cm / 8 in high, 2.5 m / 8 ft 2 in long

Flowers — April to June

Uses — Ornamental, weaving and basket-making

Grow — Buy to plant in autumn

Harvest — Long stems as required

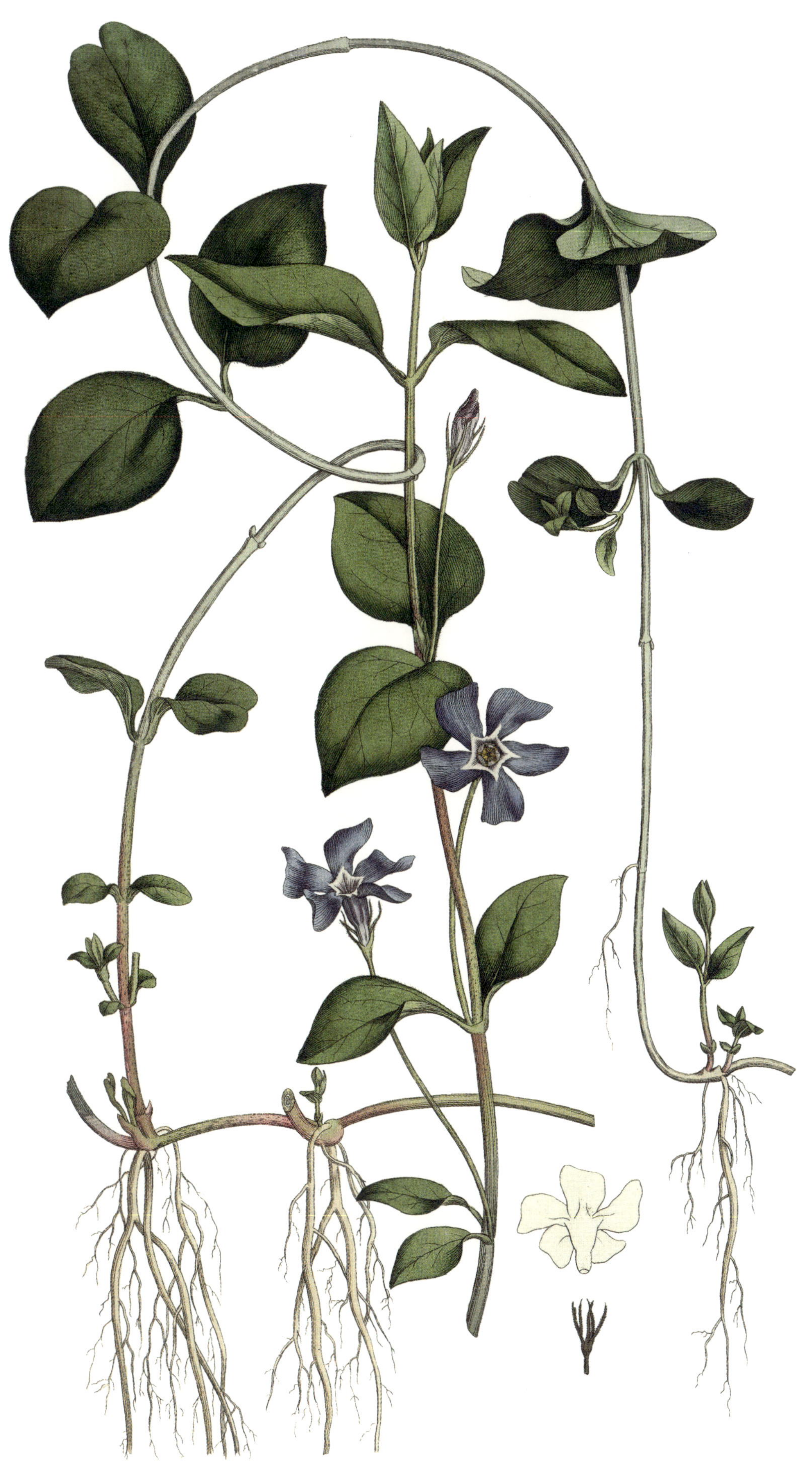

Vinca major

YELLOW FLAG

Iris pseudacorus

Glorifying even the lowliest drainage ditch, yellow flag stands proudly at water's edge, waving its big, bold flowers in the breeze like royal pennants. In fact, this heraldic plant is thought to be the inspiration for the fleur-de-lis, the emblem of French royalty. As well as being a real head-turner, yellow flag has many uses, from removing toxic metals in water systems to making impromptu toy boats. The leaves, flowers and roots produce vivid natural dyes and inks which are fun to make. Easy to grow in containers if you don't have a pond, it brings bees and a touch of regal splendour to any garden.

ANATOMY NOTES

SEEDS — The seeds can be roasted and used as a non-caffeinated coffee substitute, as was done by the people of Jersey under German occupation.

FLOWERS — An individual yellow flag flower produces more nectar per day than almost any other UK flower, a useless fact beloved by bees.

LEAVES — Children love to make the leaves into toy sailing boats by bending the leaf tip over itself and inserting it into a slit made halfway down the leaf. The tip serves as a keel poking through, and the loop is the sail.

ROOTS — Research shows that yellow flag roots soak up heavy metals, absorbing as much as 96% of the toxins – remarkably useful for water treatment. The rhizomes (the underwater roots) reproduce quickly, making the yellow flag an invasive weed in some places, but it can be contained by growing in an aquatic basket and dividing regularly.

BOTANIST'S CRAFT

CLOTHES DYE — This plant produces dyes of three different colours: yellow from the flowers, green from the leaves and black from the roots, all of which have been used in the making of Harris Tweed, the famous cloth from the Western Isles of Scotland. Place the plant material (tear up the leaves or grate the root) into a pan and cover with water. Bring to the boil and simmer for about 30 minutes until it looks bleached. Strain and dissolve a mordant, such as alum for woollens, into the liquid to help fix the colour. Add your textile to the dye bath and simmer for 20 to 30 minutes.

ESSENTIALS

Aka — Yellow iris, daggers

Family — Iridaceae (iris)

Urban habitat — Water margins

Height — 1.2 m / 3 ft 11 in

Flowers — May to July

Uses — Textile dye and ink

Grow — Buy plants to sink in shallow water in late summer

Harvest — Roots in spring, leaves and flowers in summer

At once beautiful, delicate, and singularly curious

W. Curtis

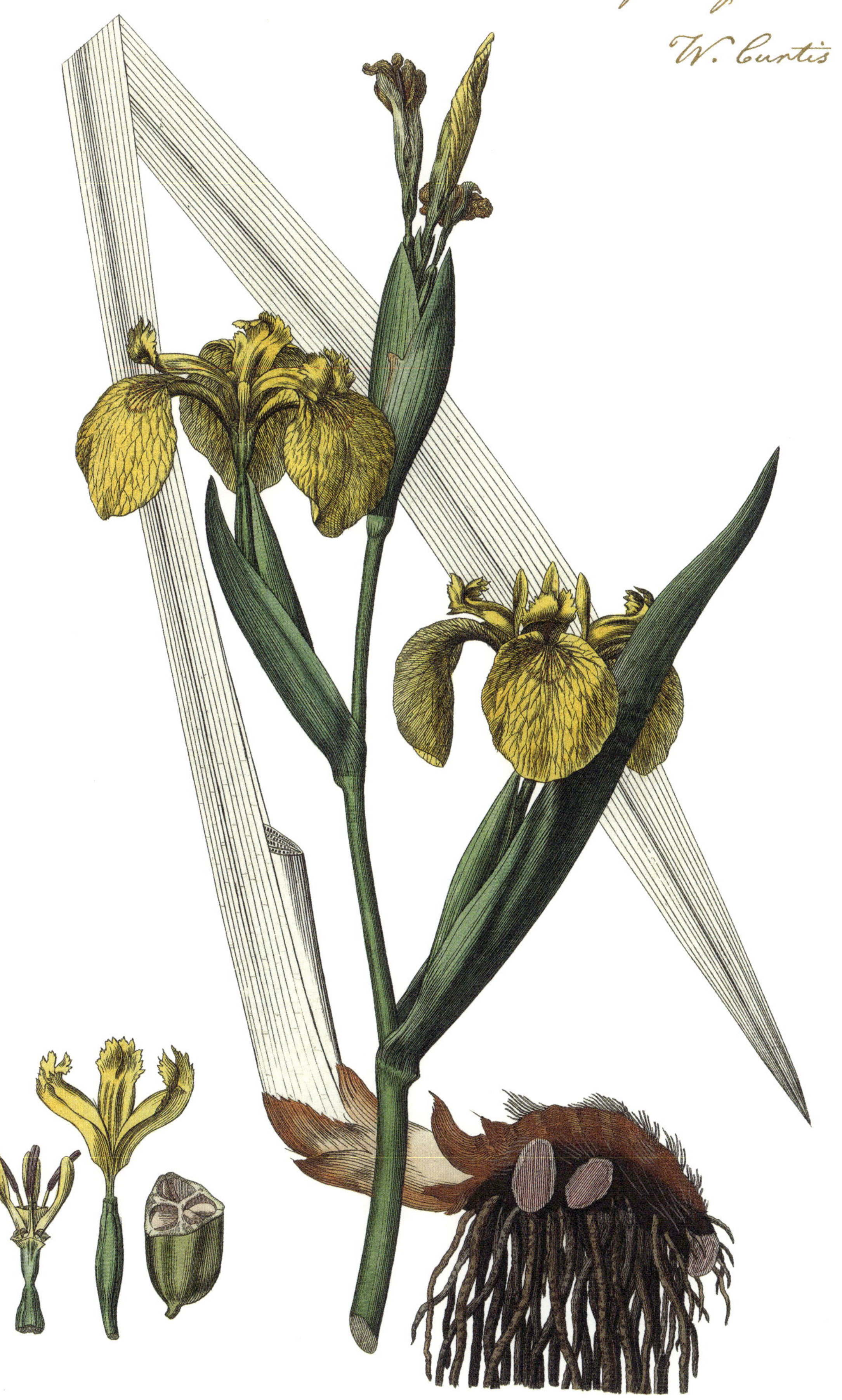

Iris pseudacorus

WILD THYME

Thymus serpyllum

This resilient little plant closely hugs the ground, withstanding the indignity of being trampled underfoot in its roadside strongholds. In summer it celebrates its hardiness in an explosion of scent and colour, forming a carpet of little lilac flowers that thicken the air with that fragrance so familiar in the kitchen. A wonderful plant for the poor soils and heat of the typical urban garden, or even to have in a balcony pot, its evergreen leaves are great freshly picked for your evening meal. And the flowers attract pollinators like honeybees too – while, when dried and popped into the wardrobe, they repel unwanted clothes moths.

ANATOMY NOTES

FLOWERS — These beautiful, but tiny, pink flowers are tricky to pick for a posy, but the Victorians still assigned a gift of one with many meanings, especially health and happiness.

LEAVES — Wild thyme leaves have long been used in many herbal remedies. They have antiseptic properties, great for soothing coughs and sore throats, and sometimes appear in commercial cough sweets and mouthwashes. A simple homemade tea also does the trick. Its essential oils are found in soaps and skincare products. The Edwardians planted whole lawns of wild thyme, so that the fine fragrance of its volatile oils would follow wherever they walked.

BOTANIST'S CRAFT

MOTH REPELLENT SACHETS — Sachets of dried wild thyme judiciously planted in your storage drawers can help to keep them away from your precious cashmeres and woollens. Harvest the leaves and flowers in July and dry by leaving them in the sun or in an open oven on its lowest setting for around 6 hours. Sew up three sides of a piece of 30 cm by 10 cm / 12 in by 4 in cotton folded in half to make a 15 cm by 10 cm / 6 in by 4 in rectangle. Fill with the herbs and tie with a ribbon. Replace the thyme every 4 months.

ESSENTIALS

Aka — Creeping thyme

Family — Lamiaceae (mint)

Urban habitat — Dry and stony ground, roadsides

Height — 5 cm / 2 in

Flowers — July to August

Uses — Edible herb, moth repellent

Grow — Sow seeds in spring

Harvest — Whole plant in early summer to dry for year-round use

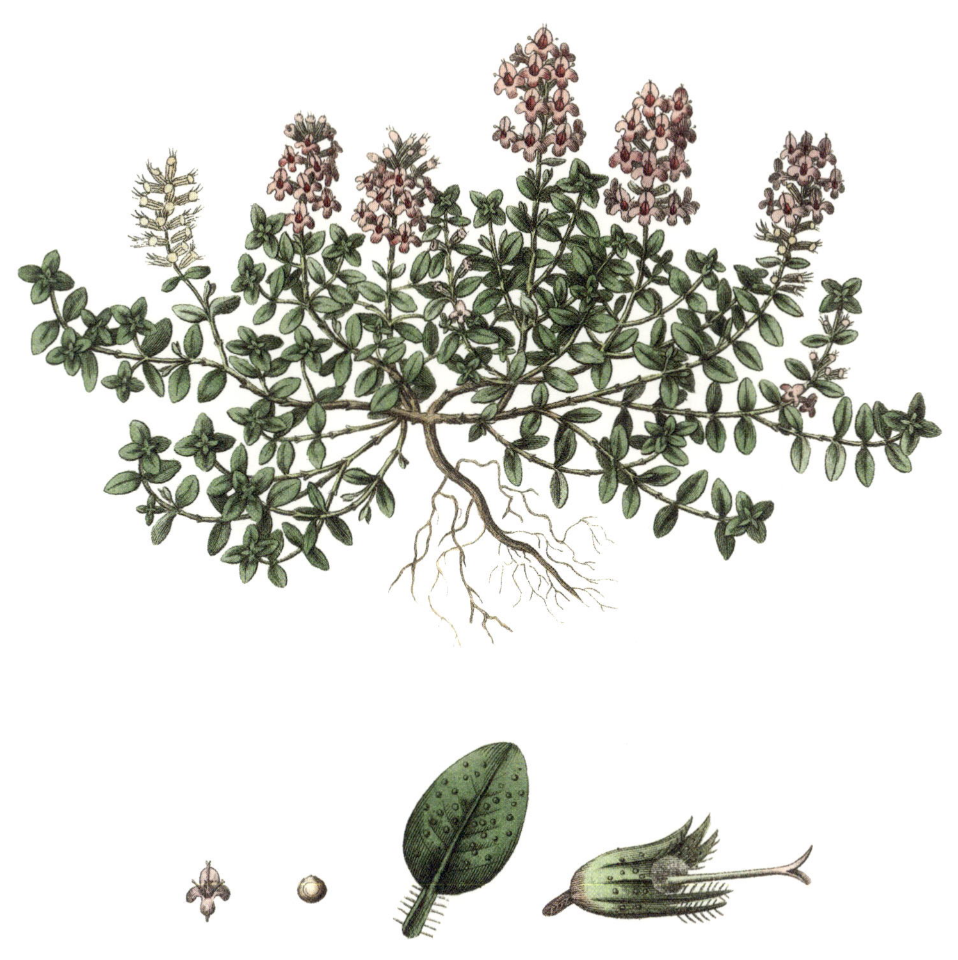

Thymus serpyllum

H E D G E B I N D W E E D

Calystegia sepium

One person's weed is another person's wildflower. But almost everyone agrees about hedge bindweed. Notoriously difficult to eradicate from gardens, it can reproduce from the smallest part of its root and its seeds can lie dormant for 30 years. Quickly twisting itself around other plants, it suffocates as it outcompetes them for light. Bindweed may fill gardeners with dread, but its virginal-white, trumpet-like flowers are both dazzling and an abundant food source for insects. Its sinewy stems make excellent instant string – small compensation for the hours spent trying to untangle them from your roses.

A N A T O M Y N O T E S

S E E D S — The seeds contain a hallucinogen related to LSD, but even hardened connoisseurs say the high isn't worth the hours of nausea that go with it.

S T E M S — Charles Darwin observed that hedge bindweed only climbs in a counter-clockwise direction, making two revolutions around another plant every 1 hour and 42 minutes.

F L O W E R S — Children love to play an old game of squeezing the green base (calyx) of the flower to 'pop' the head off, often while saying, 'Granny, granny, pop out of bed!' There are many local variants of the rhyme.

R O O T S — You can boil and eat the starchy roots, but go easy as they have laxative properties.

B O T A N I S T ' S C R A F T

C O R D — Bindweed stems snipped straight from the plant and stripped of leaves are useful as improvised string. But if you dry the stems in the sun for a day, you can plait them together to make strong cord, suitable for weaving.

E S S E N T I A L S

Aka — Ropeweed, devil's vine

Family — Convolvulaceae (morning glory)

Urban habitat — Gardens, verges, hedges and wasteland

Height — 3 m / 9 ft 11 in

Flowers — July to September

Uses — Cord, famine crop

Grow — Not advised

Harvest — As required

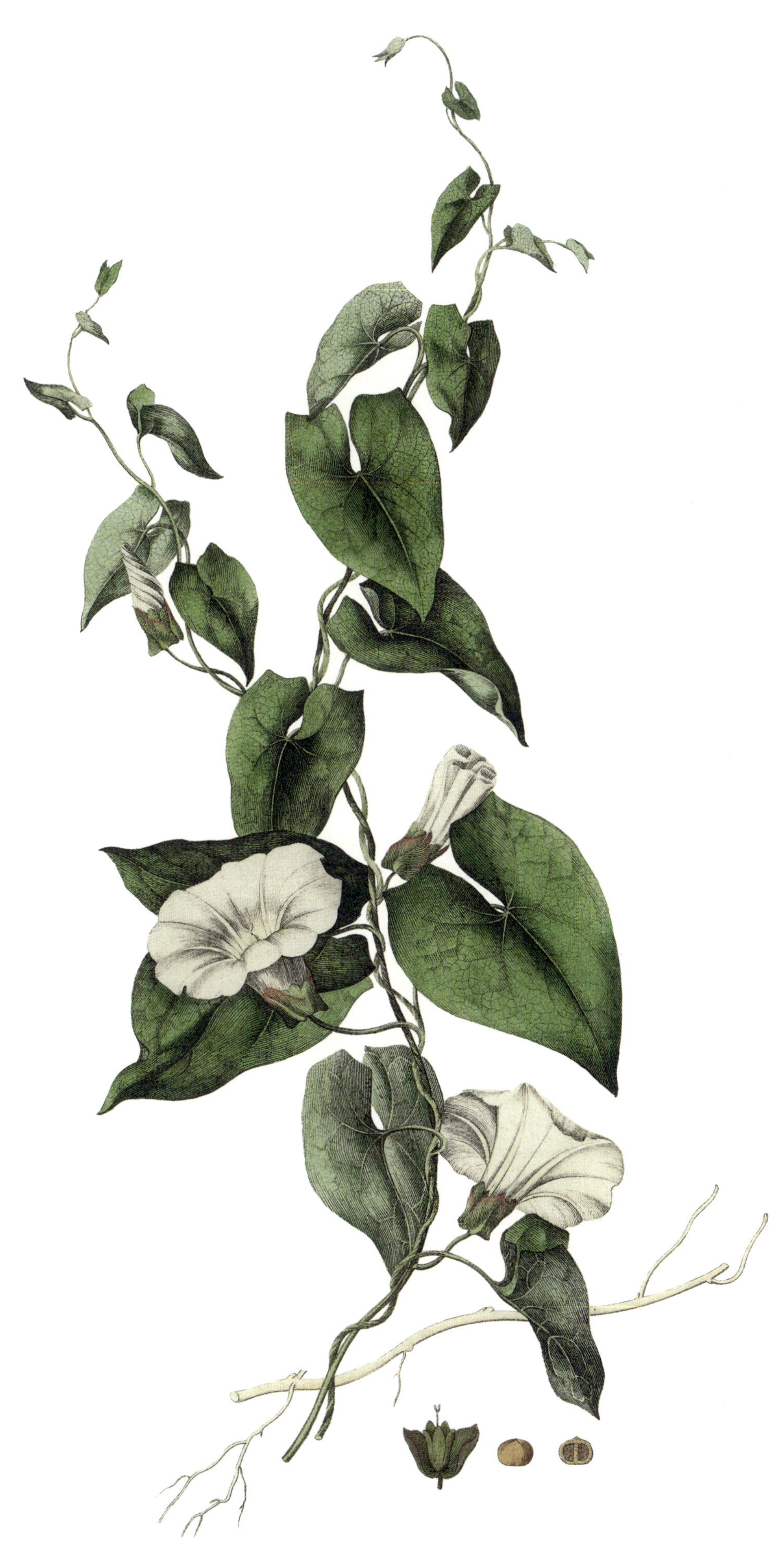

Calystegia sepium

SOAPWORT

Saponaria officinalis

Soapwort's buoyant pink blooms along some grimy ditch seem to brighten up even the grubbiest urban tableau. That's the essence of this, one of nature's most useful wild plants – though no prizes for guessing what it's used for. Full of saponins, which produce a lather in water like soap (*sapo* is Latin for soap), the plant is a natural cleanser for clothes, hair, skin and pretty much anything else. Best of all, it's very gentle, and free from the artificial foaming agents that many commercial soaps have. It's so mild, it seems our forebears used it to wash even the Bayeux Tapestry and Shroud of Turin.

ANATOMY NOTES

FLOWERS — The flowers produce most of their nectar by night to attract nocturnal pollinating moths.

LEAVES — Like the roots, the leaves can also be processed to produce cleansers. Soapwort was important for the manufacture of textiles for centuries. It was used to wash sheep's wool before shearing, then by fullers to bulk up the wool, and again to keep it clean. This is all reflected in some of its many common names: herba lanaria (wool herb) by the Romans, fuller's herb, and Bouncing Bet, an old term for a washerwoman. Herbalists also use a preparation of the leaves to relieve itchy skin.

ROOTS — The saponins are strongest in the roots, and are most potent when harvested in autumn. You can dry them for later use. Despite saponins potentially irritating the mucus membranes, the processed root extract is commercially used in the production of some tahini and halva brands.

BOTANIST'S CRAFT

SHAMPOO — Don't expect any of the grotesque frothing lather you get with a commercial shampoo. Instead, enjoy a gentle wash that leaves your hair feeling fresh, chemical-free and, after a few uses, as lustrous as you could hope for. The recipe works well with fresh or dried roots, and you can add fragrant plants to the mix too. Take 3 tablespoons of fresh cleaned and grated soapwort, or 1 tablespoon of dried root, and place into a pan, adding any fragrant plants you like. Cover with 230 ml / 8 fl oz water and simmer for 15 minutes. Allow to cool, strain and pour into a bottle.

ESSENTIALS

Aka — Bouncing Bet, fuller's herb

Family — Caryophyllaceae (carnation)

Urban habitat — By streams and waysides

Height — 80 cm / 2 ft 8 in

Flowers — July to September

Uses — Soaps and cleansers

Grow — Sow seed in spring

Harvest — Roots in autumn

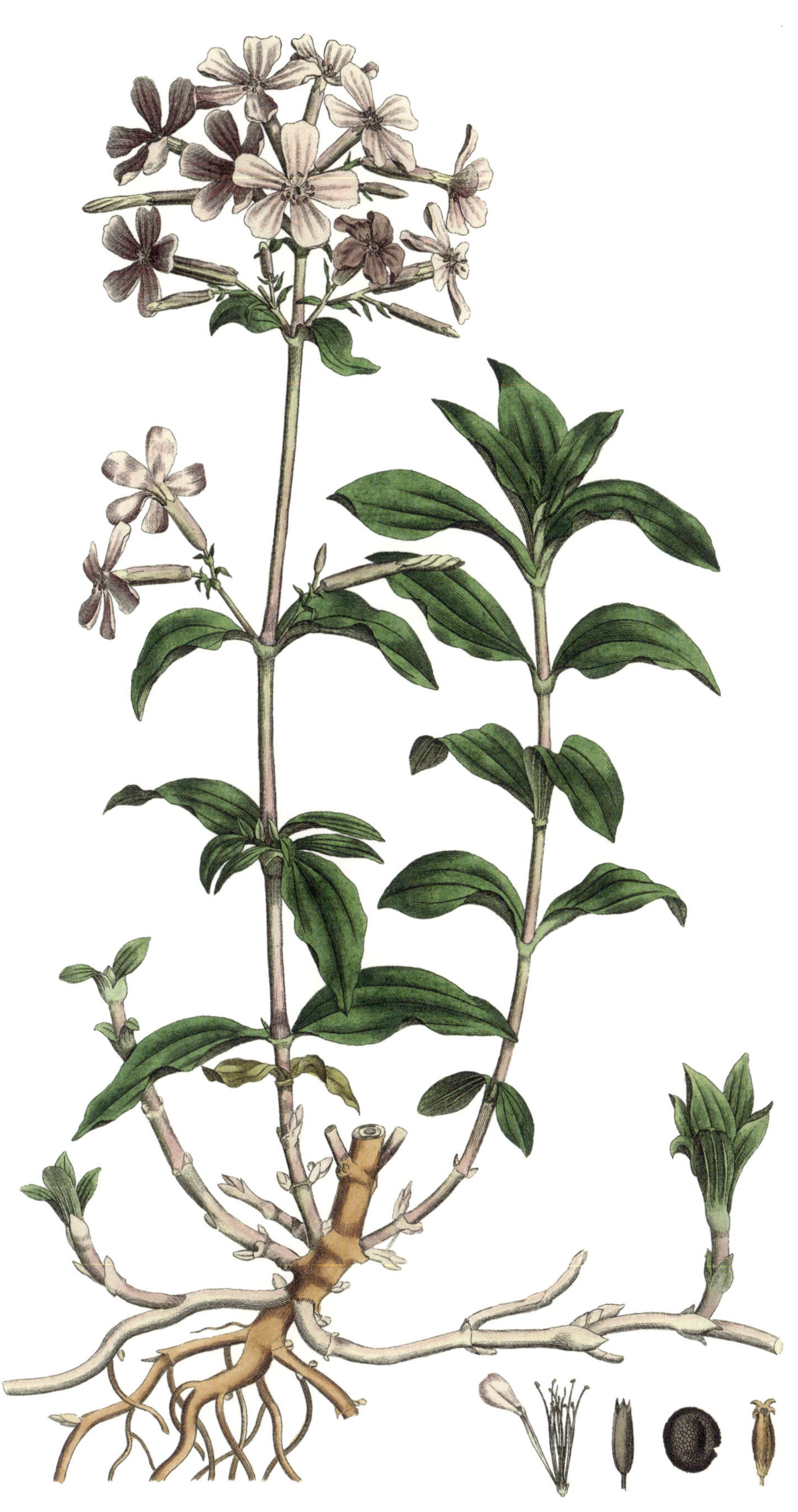

Saponaria officinalis

TRAVELLER'S JOY

Clematis vitalba

In the depths of winter when the landscape as at its barest, traveller's joy takes centre stage. The dieback of other plants reveals its large feathery seed heads laced in chain-link fences and decorating waysides like festive snowy baubles, even though they've been there since late summer. The display can last well into spring, but there's much more to this plant than a bit of fluff. Its long, twisting stems have been used to make rope and baskets since at least the Iron Age, and its flowers and seeds keep hordes of birds and insects in food. Joy indeed.

ANATOMY NOTES

FLOWERS — The flowers have a light vanilla scent. They look amazing in displays when the 'styles' (long reproductive parts of the flower) are elongated like tentacles. The flowers feed bees and hoverflies by day and many species of moth at night.

STEM — The young shoots feature in traditional Italian wild herb dishes such as *pistic* and *prebuggiún*. As the stems don't catch fire but smoulder, they can be dried and smoked like cigarettes – a former pastime of schoolboys. In some parts of the world it's an invasive species, outcompeting plants and trees, its stems strangling and suffocating as they go.

SEEDS — The fluffy part of the seed heads (which resemble the old man's beard of its name) helps the seeds spread on the wind. Many birds rely on the seeds as a food source through the winter.

BOTANIST'S CRAFT

BASKET WEAVING — The flexible stems of traveller's joy come in a variety of widths, perfect for weaving. Baskets need sturdy pieces for the structural ribs, and thinner ones for cordage to make the body, all best collected in winter. Make the frame by crossing 4 thick pieces of about 70 cm / 2 ft 4 in (one of which can be longer to be bent over at the end to make a handle) at their centres, curving in the same direction. Cut one of the ribs off at the base so you have an odd number; this will mean your layers of cord alternate as you weave your way upwards. Gradually increase the thickness of your cord as you go, trimming and tying off the rib-ends to finish.

ESSENTIALS

Aka — Old man's beard

Family — Ranunculaceae (buttercup)

Urban habitat — On trees and shrubs in hedges and roadsides

Height — 15 m / 50 ft

Flowers — July to September

Uses — Basket making, edible young shoots

Grow — Sow seed in spring

Harvest — Stems in winter for basket making

It is often made use of for arbours and bowers in gardens and pleasure-grounds

W. Curtis

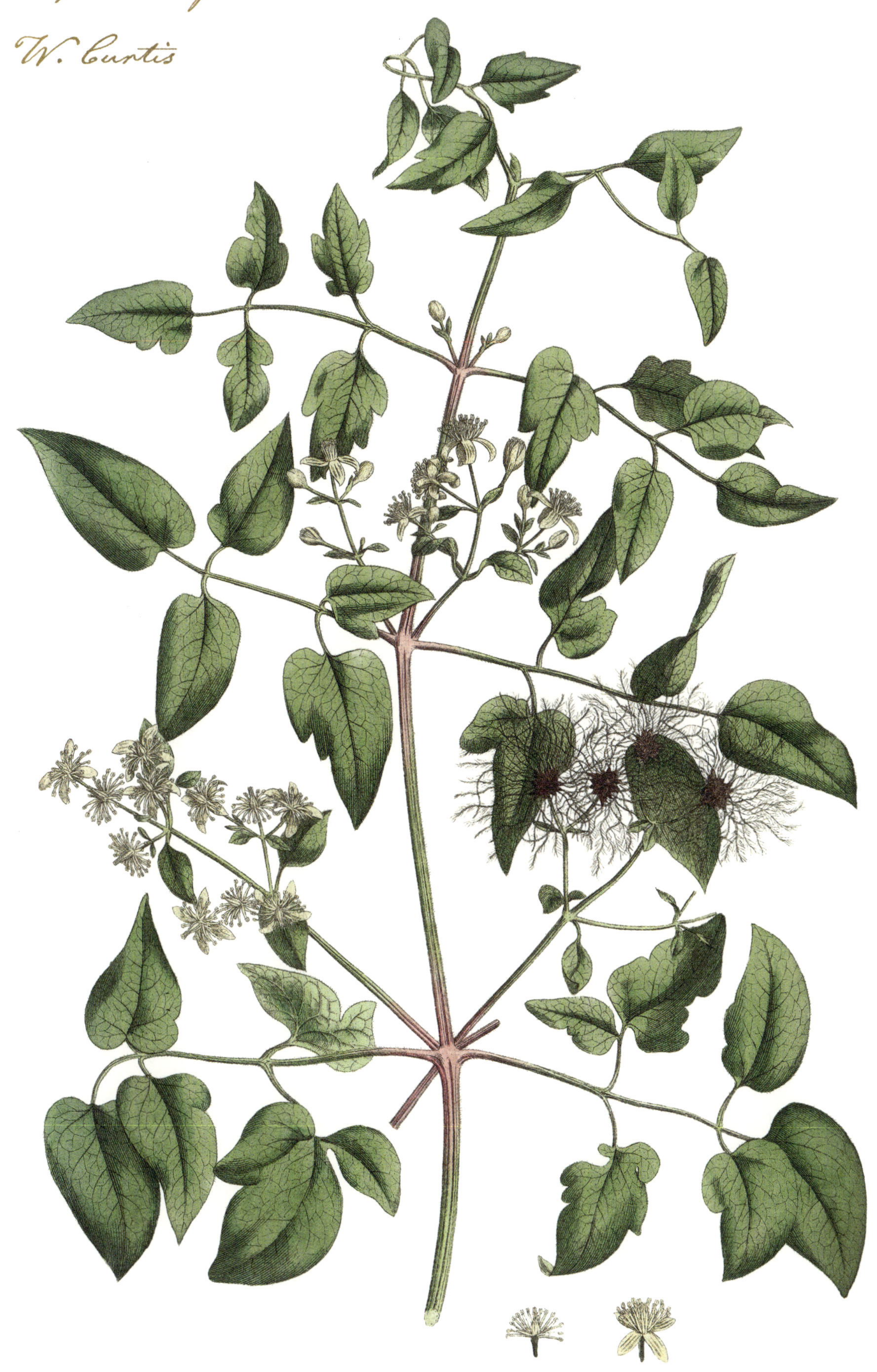

Clematis vitalba

D O G R O S E

Rosa canina

For Victorians, a gift of scented pink dog roses was a mixed message, symbolising pleasure and pain. That's because this wild rose is a bundle of contradictions. Its blossoms may be sweet, but this hard-bitten rambler will climb over just about anything that stands in its way. Its prickles catch and scratch easily, but its nutrient-rich fruits (rosehips) are used widely in skincare products. They're a delight for taste buds too, enjoyed in cakes, tarts, preserves, wines and syrups – and so packed with vitamin C that they were used in wartime to treat scurvy when citrus fruits were scarce.

A N A T O M Y N O T E S

F L O W E R S — The buds and petals taste as good as they look. Steeped in water, they create a fragrant flavouring for food and perfume. Leave some petals on the plant to get hips in the autumn.

F R U I T S — Eaten as far back as 2000 BC, rosehips are a versatile addition to any kitchen – but do strain out their little hairs which can irritate the throat. The antioxidants in the oil are great for your skin.

S T E M S — Grow your own dog rose by taking a 30 cm / 1 ft cutting just below a bud on a pencil-thick stem in autumn after the leaves fall off. Push into the ground leaving around 10 cm / 4 in showing. A new plant should take by the following autumn. Prickles on the stem curve downwards like dogs' teeth (a possible reason for the plant's name) and hook into other plants as it grows upwards. And, yes, they're prickles, adaptations of the outer layer of skin cells – not thorns, which are modified leaf stems.

R O O T S — Pliny the Elder, the Roman naturalist, thought the root of a dog rose could cure rabies bites. It seems he was wrong – there's still no cure for rabies.

E S S E N T I A L S

Aka — Witches' briar

Family — Rosaceae (rose)

Urban habitat — Roadsides, hedges, verges, scrub

Height — 5 m / 16 ft 5 in

Flowers — May to August

Uses — Skincare, food

Grow — Take a cutting in autumn

Harvest — Petals in summer, rosehips in autumn

B O T A N I S T ' S C R A F T

S K I N O I L — Place one part washed, crushed rosehips in an ovenproof dish and cover with two parts base oil, such as almond. Put in the oven at the lowest setting with the door open for 4 to 8 hours, until the oil is infused and all water moisture has evaporated. Strain through muslin making sure you remove all the irritant hairs and pour into a sterilised container.

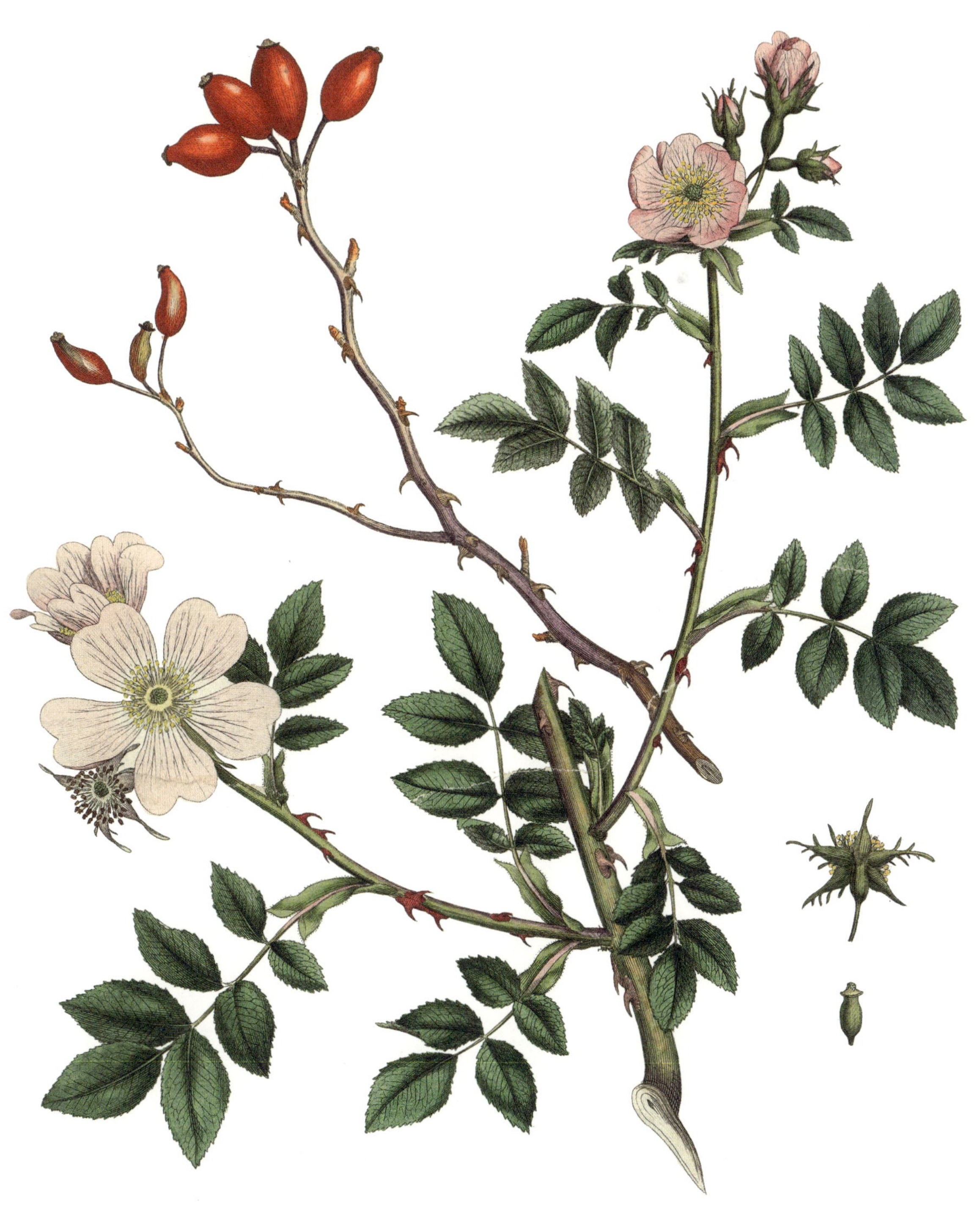

Rosa canina

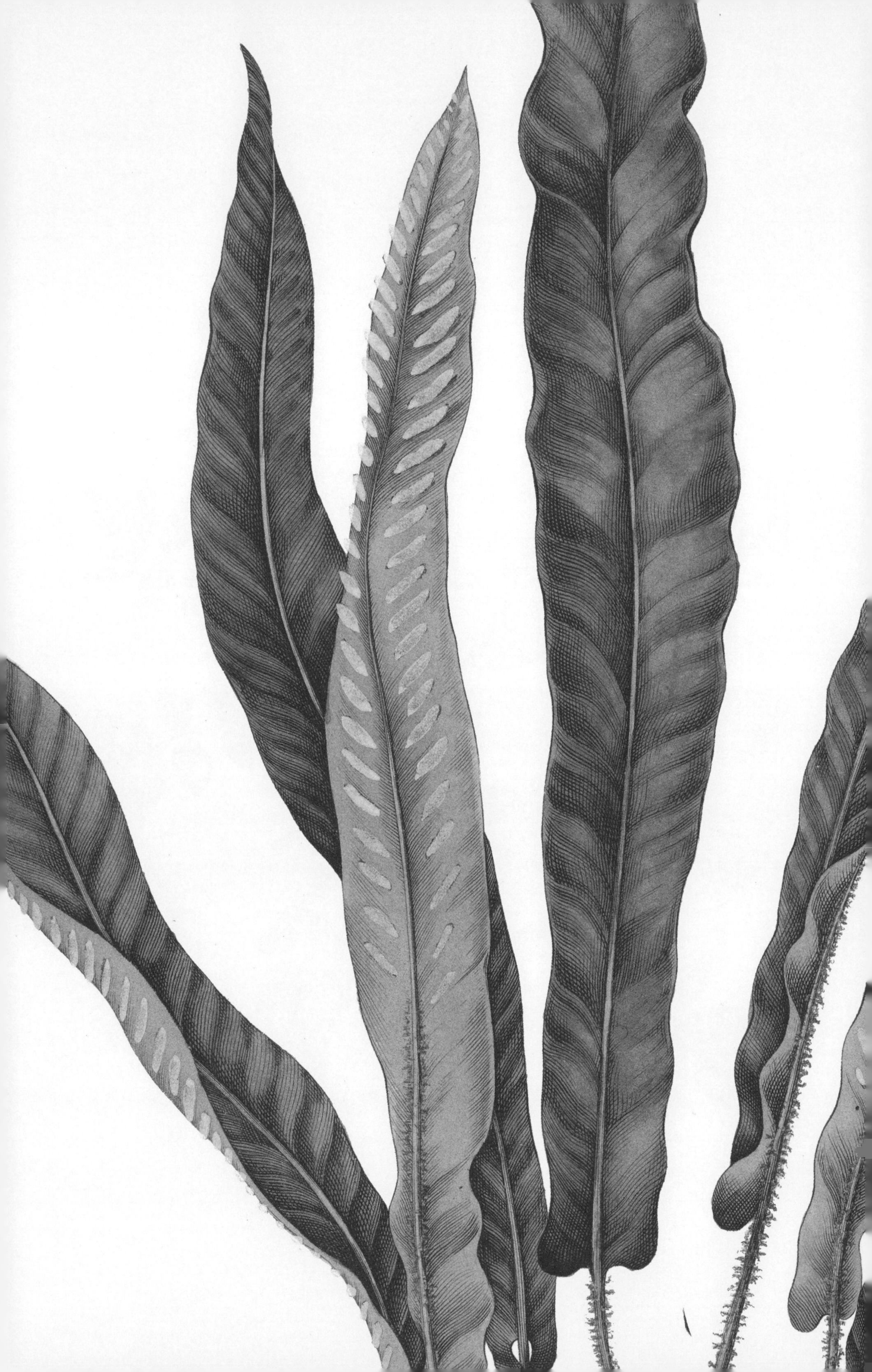

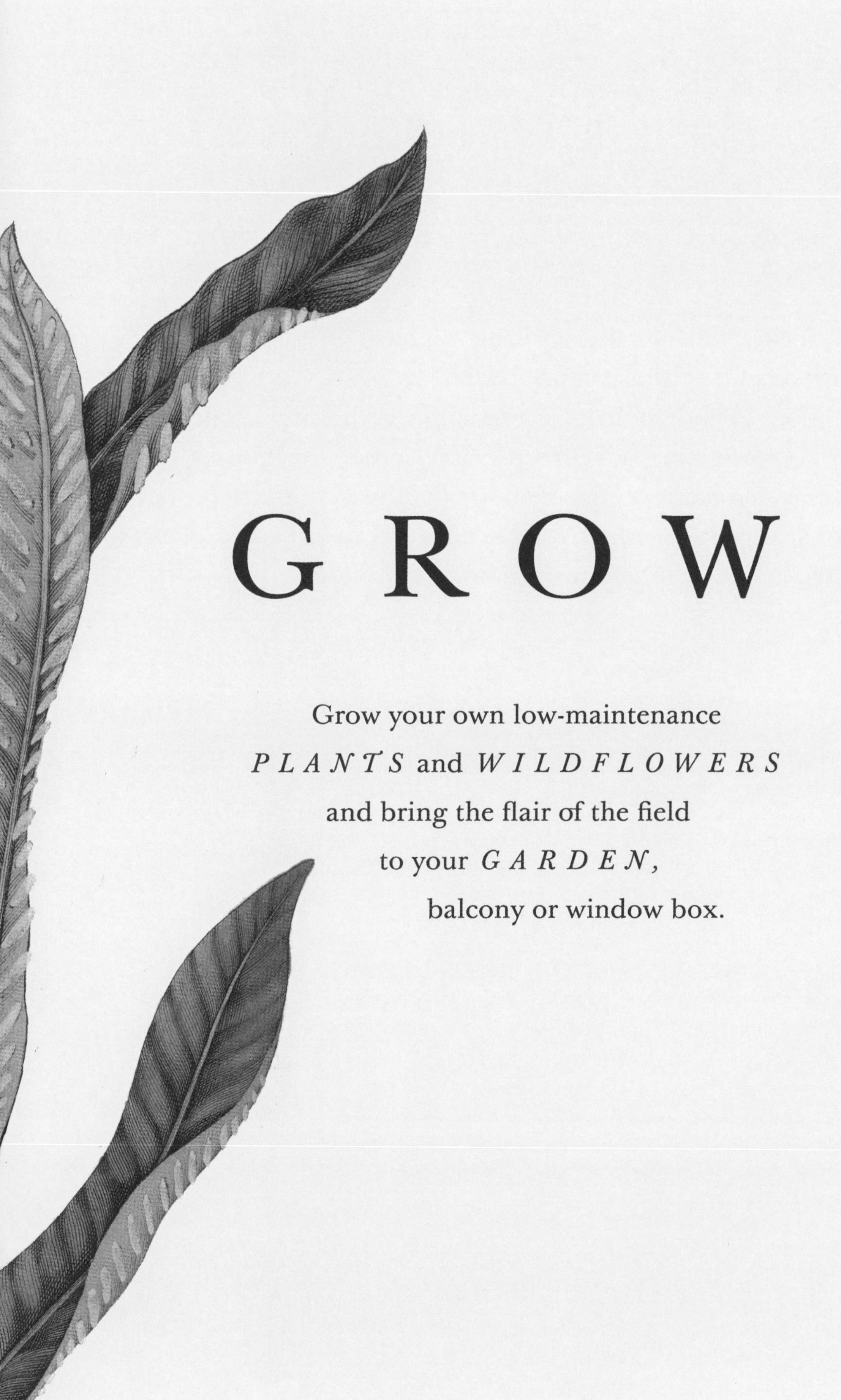

GROW

Grow your own low-maintenance *PLANTS* and *WILDFLOWERS* and bring the flair of the field to your *GARDEN*, balcony or window box.

SNAKE'S HEAD FRITILLARY

Fritillaria meleagris

Its purple, snake-skinned bells may hang sombrely like mourners' heads, but few wildflowers lift spirits as much as this exquisite fritillary. It was once plentiful in our dwindling hay meadows, but in the city you'll be lucky to spot a stand growing in undisturbed grassy corners. Why leave it to luck? You can enjoy this exotic chequered flower at home even with the minimum of outdoor space. It flourishes in beds, pots and containers, and can make the basis of stunning springtime window-box displays.

ANATOMY NOTES

FLOWERS — The flower is at its most snake-like just before it opens, when it looks just like a snake's head, coiled and ready to strike. Even its stamens inside are a bit like a forked tongue. Despite all that, its Latin name comes from its resemblance to guinea-fowl *(meleagris),* whose plumage is similarly mottled.

STEM — The stem initially droops over to allow the petals to protect its reproductive parts from rain. After pollination, it straightens so that the seedpods can stand high and catch the winds for dispersal.

LEAVES — The leaves of a single plant get larger year by year, making it easy to tell how old it is. They die back as soon as the seedpods are ripe, so be quick to have a glance.

BULB — The bulbs are remarkably poisonous and will lead to vomiting and potential cardiac arrest if ingested.

BOTANIST'S TIP

Buy and plant bulbs as early as you can in autumn and hunt out plump, white ones. Provide contrast to your display by planting snake's heads alongside the dazzling white 'Alba' variety.

ESSENTIALS

Aka — Chequered daffodil, guinea-hen flower, dead man's bell

Family — Liliaceae (lily)

Urban habitat — Undisturbed grassy areas and verges

Height — 30 cm / 1 ft

Flowers — March to May

Uses — Ornamental garden plant, cut flower

Grow — Plant bulbs in autumn

Harvest — Flowers in spring

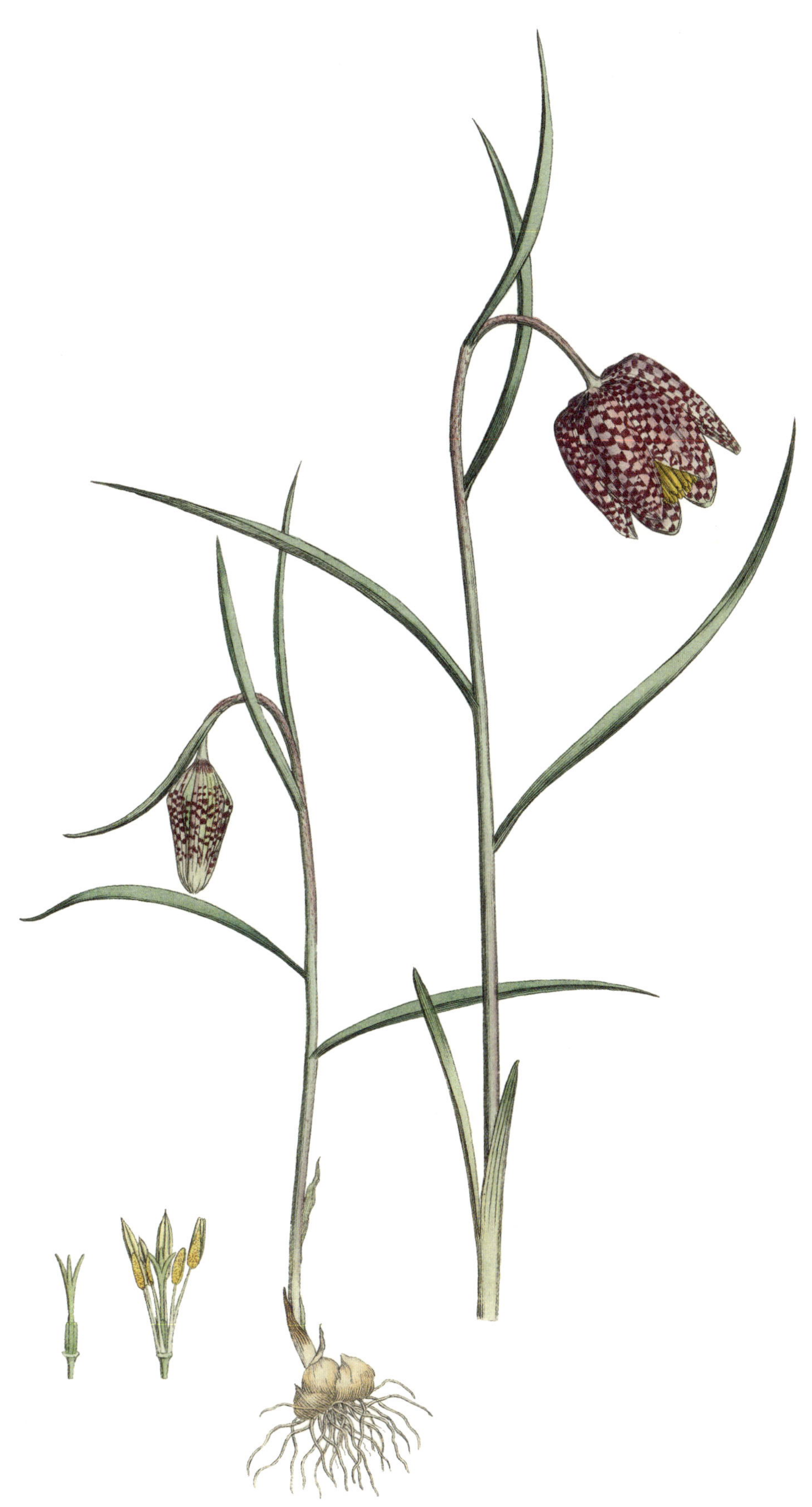

Fritillaria meleagris

C O W PARSLEY

Anthriscus sylvestris

When frothy blooms of cow parsley start shimmering by roadsides from late spring, it's a sure sign that summer warmth is on its way. Its glorious white umbels, firework explosions of tiny white flower clusters, are a familiar sight in virtually any slightly neglected grassy patch across the city. Don't let its prevalence stop you planting it in your garden; it adds wild glamour to any bed, and makes an important early food source for a huge range of insects. You can also eat the leaves and stems, which taste of parsley (funnily enough), and as you grew it yourself, there's less risk of confusing it with its deadly cousin, hemlock (see p.112).

ANATOMY NOTES

FLOWERS — There's no need to resort to expensive shop-bought umbellifer blooms when you have cow parsley to hand. Cut the flowers in May for stunning springtime bouquets.

LEAVES — You can make an attractive yellow-green dye from the leaves and stems. Harvest on a dry day in June and boil up, adding natural fibres to the strained liquid to dye.

STEM — Whether you're after cow parsley for food or dye, check the stems to make sure it's not poisonous hemlock. Cow parsley is green, occasionally with a pink tinge, whereas hemlock is mottled purple. To discourage children from picking the wrong thing, folklore has it that a mother would die if it was taken indoors – from where cow parsley gets its local name, mother-die. Once you're sure it's cow parsley, peel the stems and eat it raw or steam them with butter. They're also delicious pickled.

ROOTS — The roots taste like parsnips, but it's a lot of work for little food. Research has found the roots to be useful in treating warts and skin cancer.

BOTANIST'S TIP

Cow parsley seeds abundantly, and sowing it is a great way to get the wild form to grow at home. But there are also stunning cultivars worth trying. 'Ravenswing' bursts with frothy white heads, starkly contrasted against deep purple, almost black, stems and foliage. It's a short-lived plant, but will seed freely and herald summer in your garden for years to come.

ESSENTIALS

Aka — Wild chervil, keck, mother-die

Family — Apiaceae (carrot)

Urban habitat — Roadsides, verges, hedges and overgrown grassy areas

Height — 1 m / 3 ft 4 in

Flowers — April to June

Uses — Garden plant, animal fodder, edible leaves, cut flower, dye

Sow — In spring

Harvest — Leaves to eat in early spring (check it's cow parsley!); for dye in early summer

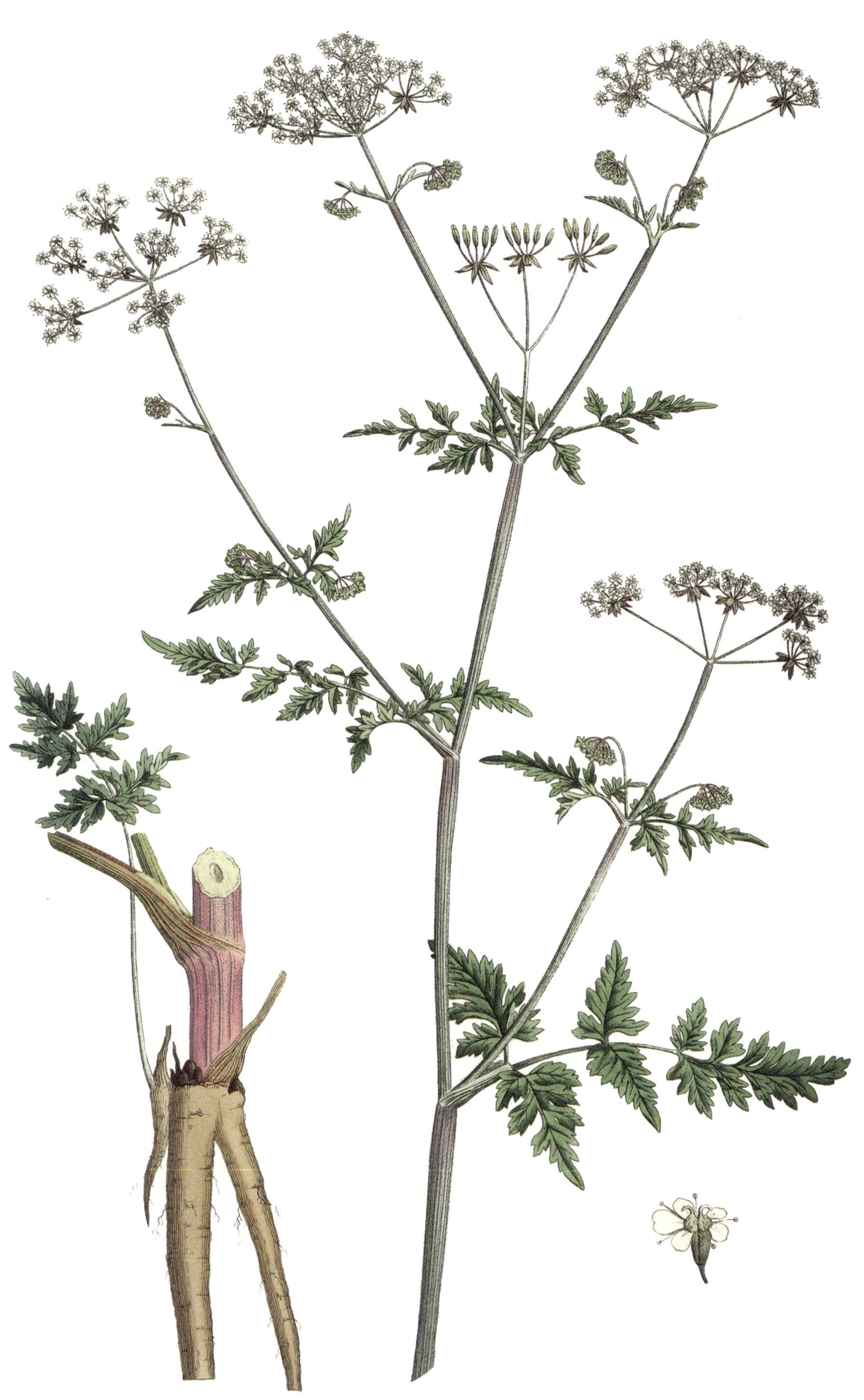

Anthriscus sylvestris

HART'S TONGUE FERN

Asplenium scolopendrium

Resembling the tongue of a deer (hart), the crinkled green leaves of this fern pop up in the dankest, shadiest corners of the city, from cracks in walls and pavements to fissures on canal bridges. It makes a perfect plant for city dwellers, as it flourishes in awkward nooks in urban gardens overshadowed by tall buildings. It's equally happy indoors as a houseplant – just make sure you keep its feet wet.

ANATOMY NOTES

SPORES — The dark brown lines of reproductive spores on the underside of the fronds, called sori, rather resemble centipedes, or *scolopendra* in Latin, which give the plant its name.

LEAVES — A syrup made from the fern's fronds is a remedy for ailments of the spleen, liver and stomach, including dysentery, according to the herbalist, Nicholas Culpeper. He also recommended a distillation of the leaves to cure passions of the heart and hiccups, while others have used it to wash greasy hair or soothe burns and scalds. In an old folk tale, Jesus lay down beside a stream and took the hart's tongue fern as a pillow. When he awoke, the story goes, he left two dark hairs down the centre of the fern. Break open a midrib (the central stem on a frond), and you will indeed find two dark lines – they're vessels carrying liquids around the plant.

BOTANIST'S TIP

As a long-time gardeners' favourite, there are many cultivars to choose from. Among the more unusual are 'Golden Queen', with its frilly yellow leaves, and 'Ramomarginatum', noted for its fraying fronds that look like seaweed.

ESSENTIALS

Aka — Burnt weed, Christ's hair

Family — Aspleniaceae (spleenwort)

Urban habitat — Shady, moist spots

Height — 30 cm / 1 ft

Flowers – Does not flower; evergreen so can be spotted year-round

Uses — Houseplant, cosmetic, medicinal

Grow — Sow spores in September

Harvest — Summer

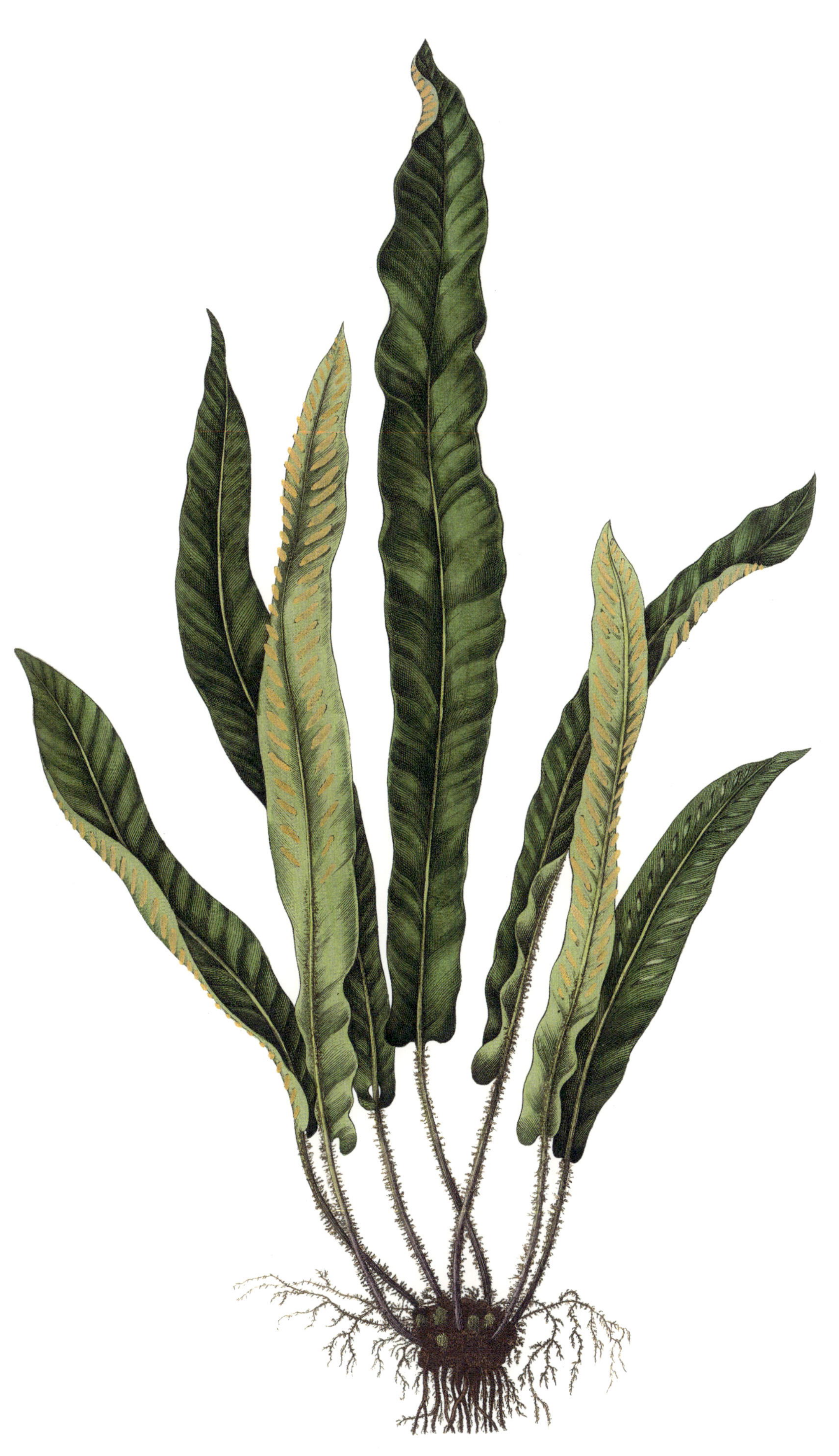

Asplenium scolopendrium

YELLOW RATTLE

Rhinanthus minor

One of the laziest plants around, yellow rattle has all the facility to live quite happily on its own, but still prefers to steal from its neighbours. It particularly loves sucking the goodness out of grasses, so it generally hasn't been hugely popular with farmers. But if wildflower meadows are your thing, yellow rattle is a veritable Robin Hood, pinching nutrients from rich pasture so that humble flowering plants can come through. As such, this is a fantastic annual to sow if you want to turn a monoculture lawn into a multicolour meadow.

ANATOMY NOTES

LEAVES — A yellow dye can be obtained from the leaves.

FLOWERS — Its beaky yellow flowers look a little like noses, which is how the plant gets its name *Rhinanthus,* from the Greek meaning 'nose flower'.

ROOTS — Once roots form in spring, they attach themselves to the roots of the most vigorous nearby plants, usually grasses and clovers, and draw water and nutrients from them.

SEEDS — Once the seedpods dry out, the seeds rattle around in their paper casings, giving the plant its common name. The seeds are food for many moth larvae, including the endangered grass rivulet.

BOTANIST'S TIP

Mow your chosen grassy patch very short in September, remove cuttings, and rake away areas to reveal soil. Sow fresh yellow rattle seed sparingly and keep it well watered. It should germinate in spring, so don't mow from then until it has flowered and released its seeds at the end of summer. When you finally cut short and rake again, you can shake out the stems to get the last seeds to fall for the next year's growth. In the following autumn, the grass should be weakened enough to sow the wildflower seeds for your meadow.

ESSENTIALS

Aka — Arctic rattlebox

Family — Orobanchaceae (broomrape)

Urban habitat — Grassy areas, cultivated land

Height — 40 cm / 1 ft 4 in

Flowers — May to July

Uses — Encourages wildflowers by suppressing grasses

Grow — Sow seed in autumn

Harvest — Seed in late summer for sowing fresh

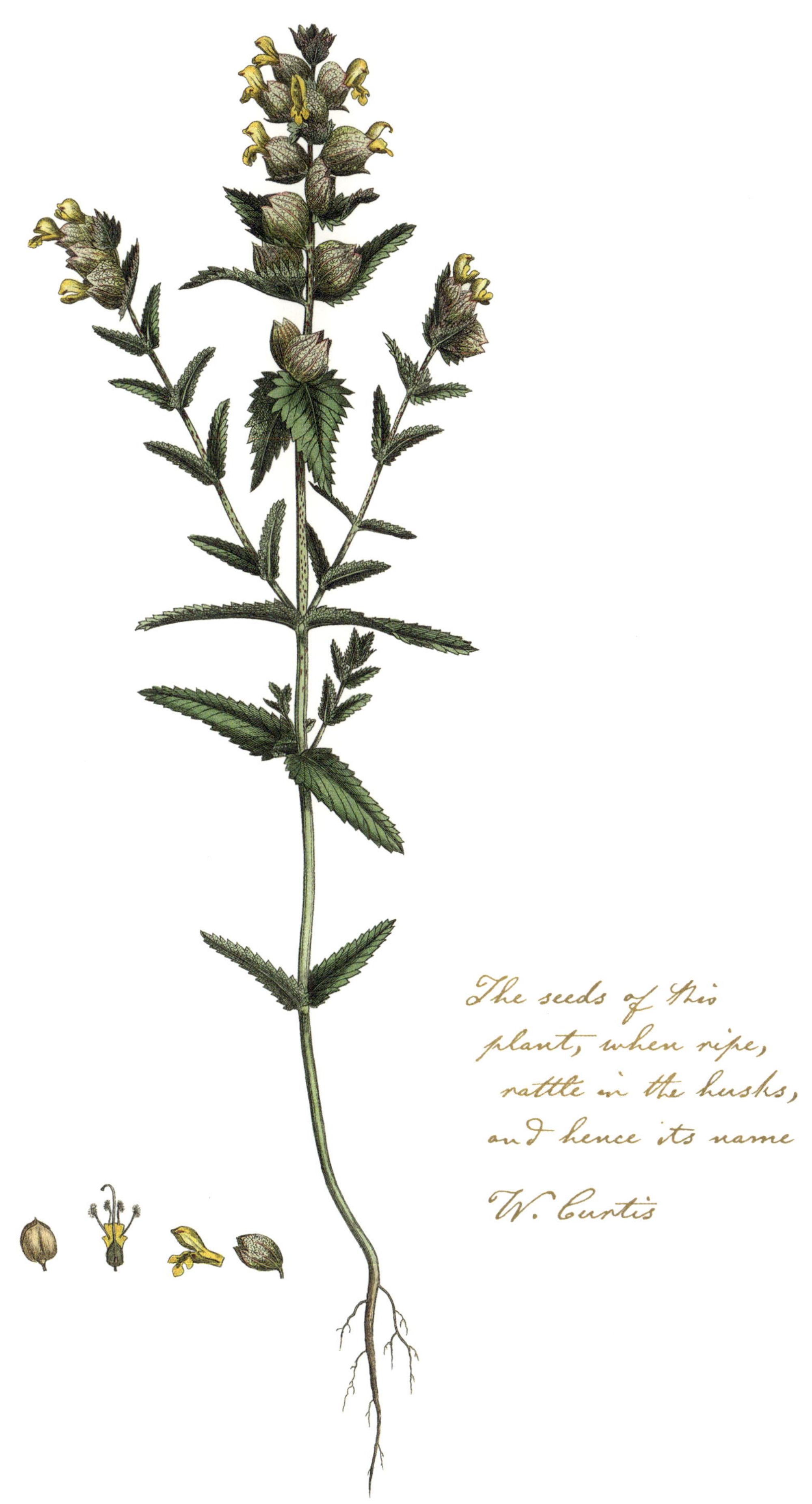

Rhinanthus minor

W A T E R V I O L E T

Hottonia palustris

Feathered whorls of luscious green swirl around the waterline wherever this aquatic delight puts down its roots. These dense, fernlike leaves oxygenate standing water, nurturing numerous insects, larvae and fish fry, and providing makeshift sunbeds for basking dragonflies. In late spring, delicate spires of pale pink flowers emerge from the surface, blinking into the daylight with their yellow-centred eyes. A hardy plant that virtually looks after itself, it's an excellent choice for shallow ponds, boggy patches and containers.

A N A T O M Y N O T E S

R O O T S — You'll find two types of root on the water violet. Sturdier ones at its base bury into the mud to stabilise the plant, while very fine ones further up the stem wave around in the water to take up nutrients.

F L O W E R S — The English homeopath Dr Edward Bach (of Bach Flower Remedies) believed his essences of water violet flowers helped lonely, anti-social people connect with others.

L E A V E S — The submerged leaves produce oxygen during the day, enriching and cleaning the water to support aquatic life and prevent the build-up of harmful algae.

P E T I O L E — As an adaptation to living in the water, the leaf stems (petioles) contain specialised cells that allow easy circulation of oxygen around the plant, especially to the roots.

B O T A N I S T ' S T I P

Water violets like shallow water in full sun, with some soil below for their roots, a habitat you can recreate remarkably well in container ponds. Simply pop a few stems into the container, and they will soon send out roots and new leaves.

E S S E N T I A L S

Aka — Featherfoil

Family — Primulaceae (primrose)

Urban habitat — Disused canals, ponds, bogs and ditches

Height — 30 cm / 1 ft

Flowers — May to June

Uses — Oxygenating aquatic garden plant

Grow — Divide existing plants in spring

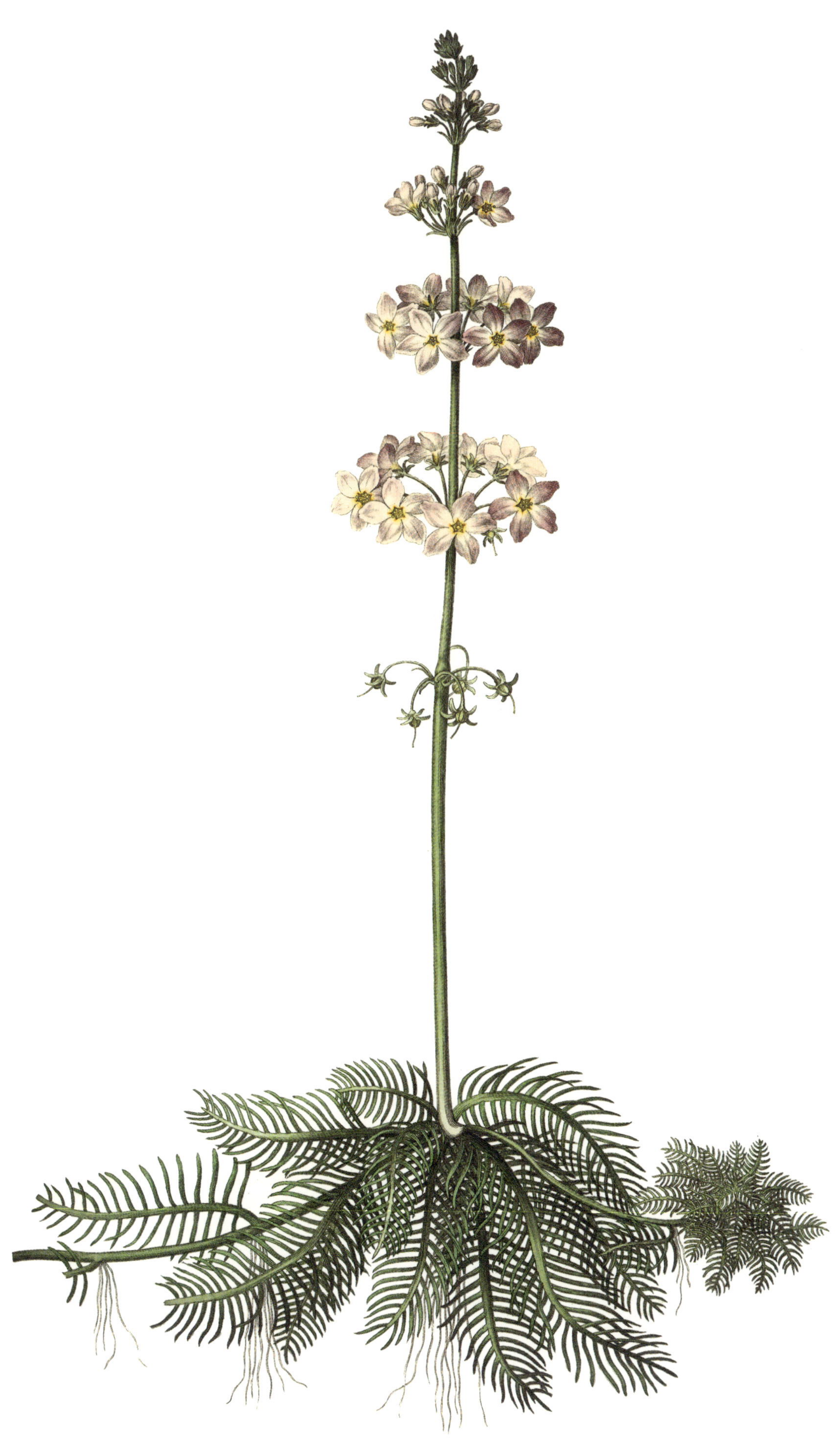

Hottonia palustris

BASTARD BALM

Melittis melissophyllum

With dignified, nectar-rich flowers that seem to unfurl miniature pink carpets to visiting bees, this threatened plant should be far more valued than its ignoble name implies. Favouring West Country woodlands in the wild, discerning gardeners have long grown it in urban gardens, as it never fails to bring a touch of sylvan charm. It thrives in broken light, including up against walls, and casts a crisp woody scent that will whisk you far from the city straight to a forest glade. Plant it if only to help make this rare treat less elusive about town – and because bees adore it too.

ANATOMY NOTES

LEAVES — Its leaves closely resemble those of its relative lemon balm, once known simply as balm. With fewer medicinal and culinary qualities than its cousin, it was referred to as 'bastard' balm to set them apart. Drying the leaves enhances and improves their taste and fragrance, making them an interesting and unusual herb to add to dishes. As a tea, it soothes an upset stomach and an anxious soul. Essentials oils from the leaves have been shown to have potential antibacterial and antioxidant effects, good for healing wounds.

FLOWERS — The attractive pink stripe on lower petal of the flower serves as a landing pad for bees, guiding them to the nectar, pollinating as they go. A circular gland within the base of the flower secretes large quantities of nectar, making it one of their favourites. In fact, its relationship with bees is so ingrained its Latin name literally means 'bee-leafed bee plant' (so much nicer than 'bastard').

BOTANIST'S TIP

Bastard balm is happiest sheltered by dappled shade, in moist, well-drained, humus-rich soil. An excellent cultivar to try is 'Royal Velvet Distinction' for its elegant orchid-like flowers tinged with a shock of pink-purple.

ESSENTIALS

Family — Lamiaceae (mint)

Urban habitat — Shady, undisturbed spots

Height — 50 cm / 1 ft 8 in

Flowers — May to August

Uses — Food for bees and pollinators, herbal tea

Grow — Divide existing plants in autumn

Harvest — Leaves in spring

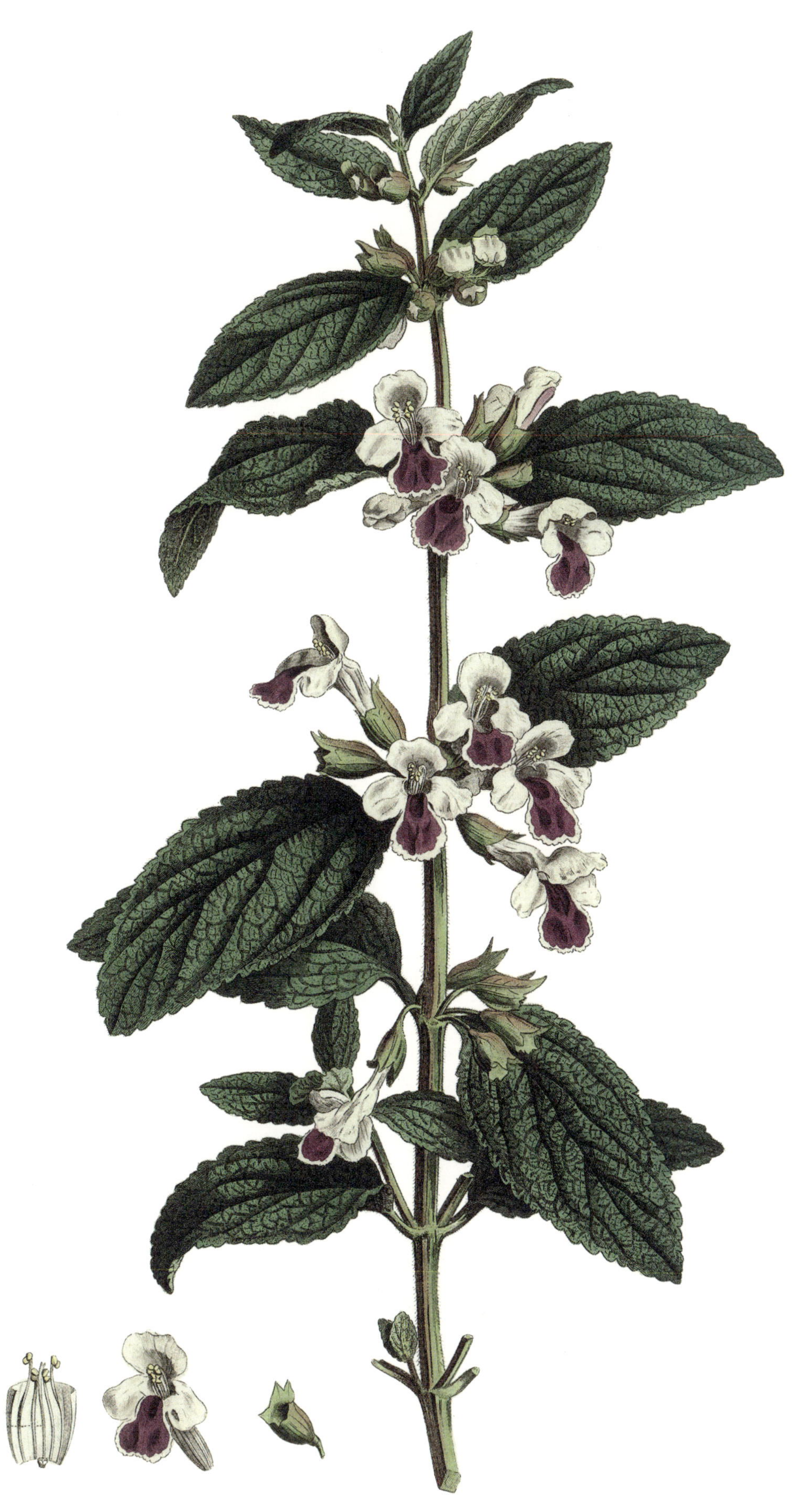

Melittis melissophyllum

BEE ORCHID

Ophrys apifera

Stumbling across this, one of Britain's most beautiful native orchids, on a grassy roadside or railway cutting is very special indeed. You're in the presence of the queen of deceit. This orchid's striking velvety flower perfectly mimics a certain female bee, even exuding its scent, to lure in hapless males. The hopefuls arrive, pollinating the plant, but leave empty-handed, finding neither mate nor nectar. The last laugh, however, is on the orchid. The particular bee that she impersonates doesn't live in the UK, so she has to make do with self-pollination. So, give this tricksy, but stunning, orchid a helping hand and grow her in your garden.

ANATOMY NOTES

FLOWERS — Involving scent, sight and feel, the bee orchid's multisensory deception is elaborate. The flower's lower lip not only has the colouration and smell of a female bee, but also its hairiness. The sepals even look like wings.

POLLINIA — For want of the right bee to pollinate it, this orchid can do the work itself. Its pollinia, dangling balls of pollen, hang down over the mouth of the flower and are blown around by the wind. Eventually, they catch the stigma inside the flower, resulting in self-pollination.

ROOTS — Orchid roots can be processed into a flour called salep, which is so nutritionally rich that an ounce is enough to sustain an adult for a day. Unfortunately, as the bee orchid's roots are rather meagre, it's not worth cultivating or harvesting them for this purpose.

BOTANIST'S TIP

Get your bee orchids from a reputable nursery, not the wild. This is a great plant to grow if you aren't a fan of mowing. Let your grass grow long from April to September and plant the orchids into it in the autumn.

ESSENTIALS

Family — Orchidaceae (orchid)

Urban habitat — Gardens, chalky soils, verges

Height — 30 cm / 1 ft

Flowers — June to July

Uses — Garden ornamental, flour

Grow — Buy plants and establish in autumn

Much sought after by florists, whose curiosity often tempts them to exceed the bounds of moderation, rooting up all they find

W. Curtis

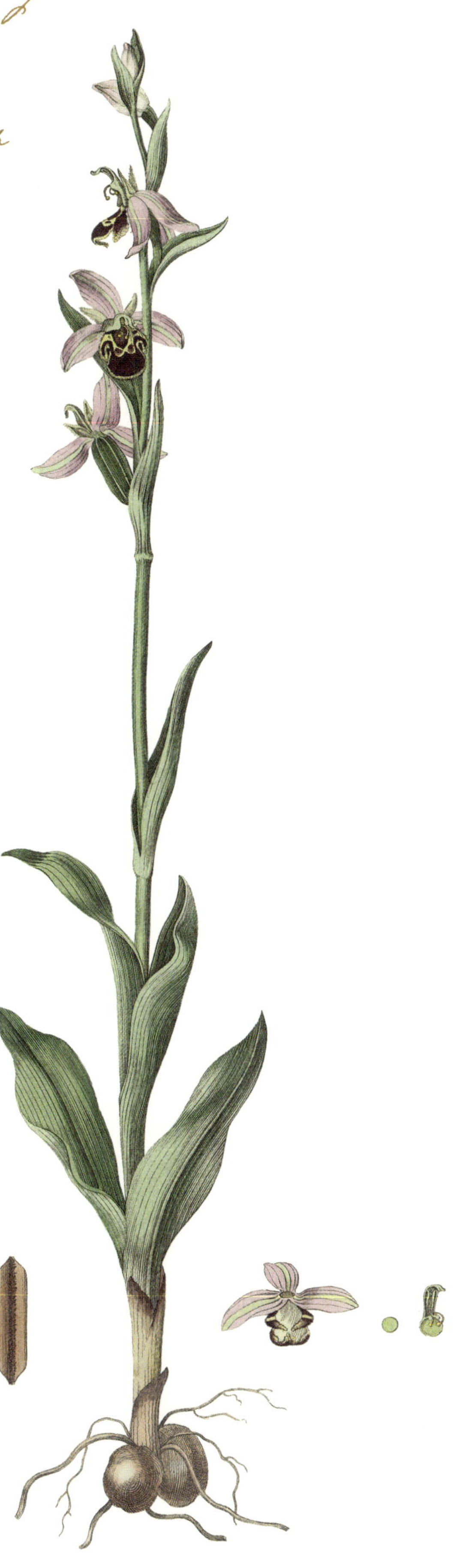

Ophrys apifera

HONEYSUCKLE

Lonicera periclymenum

Happy even in the shadow of city skyscrapers, honeysuckle might go unnoticed if it didn't fill whole streets with its delicious scent on summer evenings. A longstanding national favourite, its fragrant pinky-yellow flowers, charged with sweet nectar, don't only endear it to gardeners, children, writers and poets. Its dense canopy and autumn berries provide board and lodging for a variety of nesting birds and mice, while its juicy flowers support a long list of pollinating insects, including the hawk-moth and rare white admiral butterfly. Champion of the urban ecosystem and saviour of the shady garden, no wonder everyone loves honeysuckle.

ANATOMY NOTES

LEAVES — The leaves have long been used in herbal medicines for many ailments, from aches and pains to flus and fevers. Honeysuckle contains high concentrations of salicylic acid, the principal component of aspirin.

STEM — Honeysuckle stems twine clockwise around other plants. If they become stout, they take on corkscrew twists much loved by walking stick makers. Honeysuckle's entwinements have often reminded poets of a lovers' embrace. In *A Midsummer Night's Dream* Titania woos Bottom: 'Sleep thou, and I will wind thee in my arms... So doth the woodbine the sweet honeysuckle gently entwist.' (Woodbine here is thought to refer to bindweed!)

FLOWERS — You can do more with the flowers than just suck the honey-flavoured nectar straight through the base of the flower – something that never fails to delight children. Gather a handful to flavour water, tea and cocktails. Honeysuckle syrup is great for a scratchy cough.

ESSENTIALS

Aka — Woodbine

Family — Caprifoliaceae (honeysuckle)

Urban habitat — Shady hedges and woods

Height — 6 m / 20 ft

Flowers — June to September

Uses — Medicinal, culinary, scented garden plant, habitat and food source for wildlife

Grow — Sow seed in spring

BOTANIST'S TIP

Plant this climber with its roots firmly in the shade. The flowers like the sun, but it's important to keep the roots cool in moist soil. Give it something to cling to – either another plant or put up a structure or wiring.

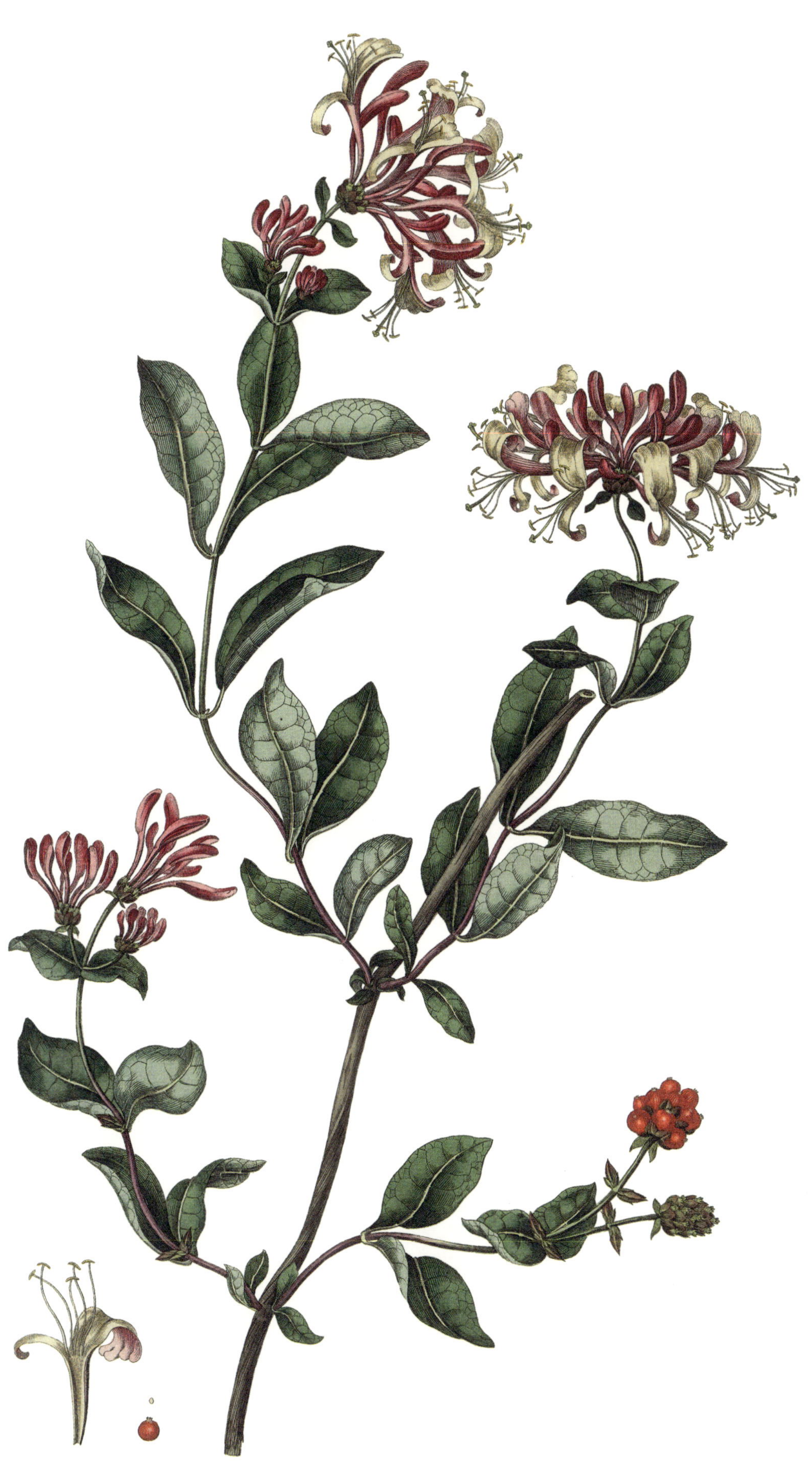

Lonicera periclymenum

HOUSELEEK

Sempervivum tectorum

If you describe yourself as more of a butterfingers than a green-fingers, then the houseleek is for you. *Sempervivum* means 'ever living' – this one is hard to kill. Originating from harsh mountain environments, it'll treat a sunny perch on a cracked old wall in a parched city garden as a comparative paradise. With you barely lifting a (buttered) finger, it will bloom into dainty pink flowers and spread, growing new fleshy rosettes of succulent leaves into whatever space you give it. And according to ancient lore, if grown on the roof, it will guard your house against fire, lightning and illness. Win-win!

ANATOMY NOTES

FLOWERS — Once a rosette flowers, it dies off, but is quickly replaced by other rosettes, called offsets.

LEAVES — The leaves have historically been used to cool and soothe such ailments as ulcers, tongue fissures, piles and eye inflammations; but its easiest everyday use is for burns, much like aloe vera. Break the leaf, remove the outer layer and apply to the affected area of skin. Young leaves are edible and add a moist yet crunchy element to salads, a kind of substitute cucumber.

OFFSETS — The houseleek spreads by producing lots of clones, a process that has earned it the common name 'hen-and-chicks'. The main rosette is the hen, putting out lots of smaller offsets, the chicks. Offsets are easily removed to plant in new positions, allowing you to cover large surfaces with ease.

BOTANIST'S TIP

For a really vibrant-red leaved houseleek, seek out the cultivar 'Rubin'. Like all houseleeks, it will eventually form a large mat of rosettes. You can snip new rosettes off and plant them in decorative pots – making perfect little gifts.

ESSENTIALS

Aka — Jupiter's eye, hen-and-chicks, thunder plant

Family — Crassulaceae (stonecrop)

Urban habitat — Sunny walls, rocky places and roofs

Height — 20 cm / 8 in

Flowers — June to July

Uses — Ornamental plant, burns treatment

Grow — Sow seed in spring

Harvest — Take leaves as needed for burns

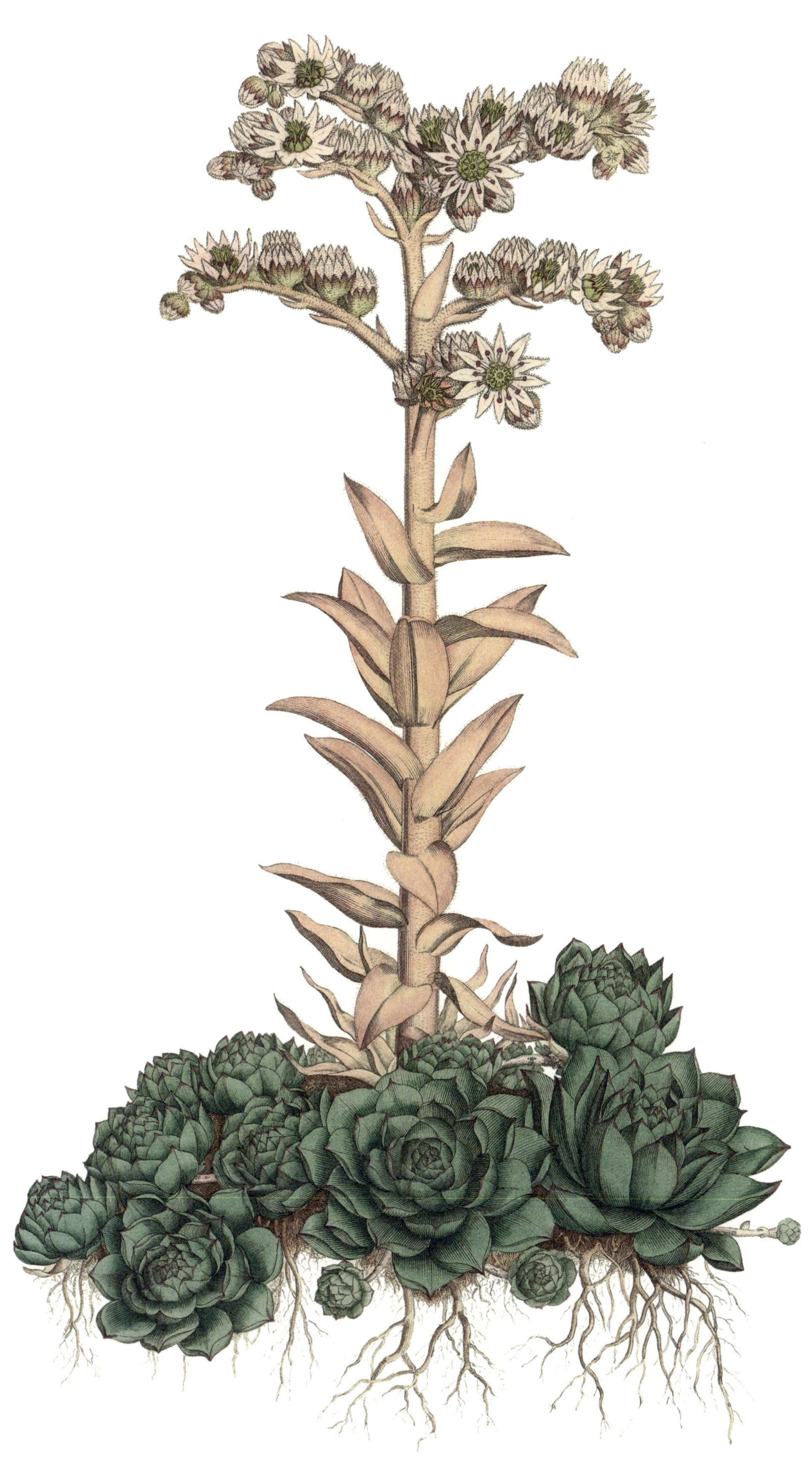

Sempervivum tectorum

TOADFLAX

Linaria vulgaris

Turning up in unloved corners – whether a dusty roadside or the broken tarmac of an old tennis court – toadflax's buoyant spikes of buttery yellow flowers look good enough for a florist's window. It's similarly unfussy in the garden, colonising those difficult spots you've given up on, and delivering its snapdragon-type blooms to grateful bees, both bumble and honey, well into autumn. Cut flowers last for ages indoors, and if you feel like putting your herbalist's hat on, it has healing powers, not least as an astringent for cleansing wounds.

ANATOMY NOTES

LEAVES — The leaves were used to make a laxative tea, or an ointment for skin complaints. One 17th-century instruction describes placing the leaves under bare feet to ward off fevers. More reliably, they can be used to make a poultice for wounds.

FLOWERS — The flowers are designed for bees; the egg-yolk orange streak on the lip guides them in, allowing a gateway within the flower to open just wide enough to reveal the nectar. The buzzing visitor gets covered in pollen as it drinks with its tongue. The blooms can produce a yellow dye, or, boiled in milk, they're said to repel flies. An infusion of flowers can be made into an antiseptic wash for cuts and scrapes. Children love to squeeze the flower lips to make them squeak.

BOTANIST'S TIP

In the garden, toadflax brings warmth and colour to any grassy wildflower area or to those awkward dry patches where other plants struggle. Sow into pots in early spring for planting out in late spring. Once they're established, you can dig up and divide them each spring to share with friends. Deadhead them before seeding to stop them getting out of hand.

ESSENTIALS

Aka — Common toadflax, yellow toadflax, butter and eggs, wild snapdragon

Family — Plantaginaceae (plantain)

Urban habitat — Hedges, ditches, waysides, waste ground

Height — 50 cm / 1 ft 8 in

Flowers — June to October

Uses — Wound wash, laxative tea, fly repellent

Grow — Sow seed in spring

Harvest — As flowers form

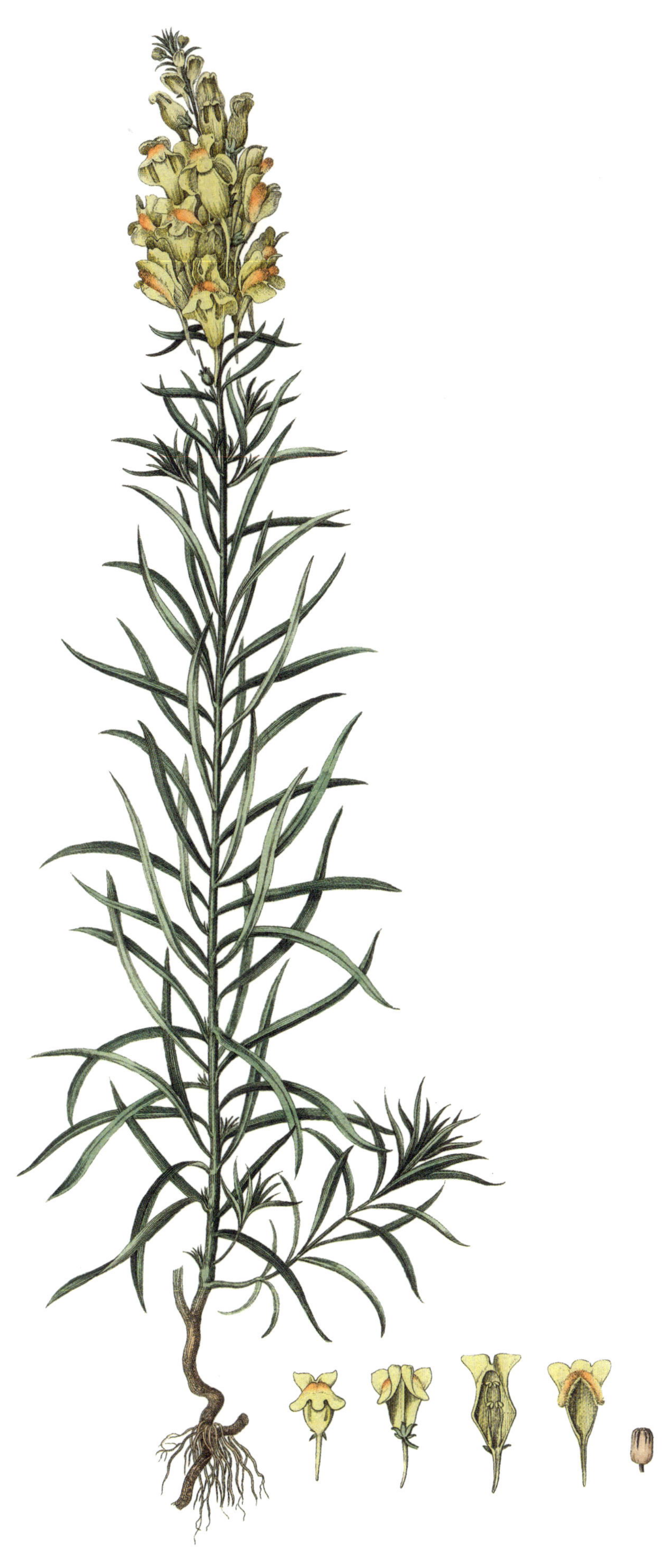

Linaria vulgaris

TEASEL

Dipsacus fullonum

Growing to the height of the average basketball player, a stand of teasels along a canal-side is an impressive sight. In summertime, it's topped off with thistle-like domes, primed with row upon row of tiny purple flowers that bees and butterflies go crazy for. In autumn the flowers dry and shed, leaving a satisfyingly bristly seed head, particularly attractive to hungry goldfinches – and florists. Good for wildlife and a stunner both in bloom and dried, whether in a vase or in a bed, this is a safe bet for any cottage garden.

ANATOMY NOTES

FLOWERS — Giving the flowers and seed heads in a bouquet is supposed to symbolise jealousy or misanthropy, probably something to do with the spikiness.

LEAVES — The leaves and other aerial parts of the plant can be turned into a blue dye that isn't a bad substitute for indigo.

SEED HEADS — The spiny seed heads of *Dipsacus* have been used by fullers since ancient times to 'teasel' woollens, to brush and comb them clean and free of knots, and to lift the nap of the cloth.

STEM — The stems are also spiky, and ward off grazing herbivores. Look closely to spot another pest defence: little water-filled pools collect where the leaves meet the stem to stop insects climbing up to suck sap from the tenderest parts of the plant. Another theory is that teasels actually gain nutrients from the drowned insects.

BOTANIST'S TIP

Teasel is easily grown from seed in the garden: just sow and rake in to cover with soil. They're biennial, so they won't produce a flowerhead until the second year of growth, just spiky leaves. They will then seed freely, providing a feast for the birds and free seedlings for you – so many, in fact, that it may be worth weeding out a few to keep it under control.

ESSENTIALS

Aka — Common teasel

Family — Caprifoliaceae (honeysuckle)

Urban habitat — Sunny margins, especially near water

Height — 2 m / 6 ft 7 in

Flowers — July to August

Uses — Comb in textile processing, ornamental

Grow — Sow seed in autumn

Harvest — Seeds heads at the end of summer for drying

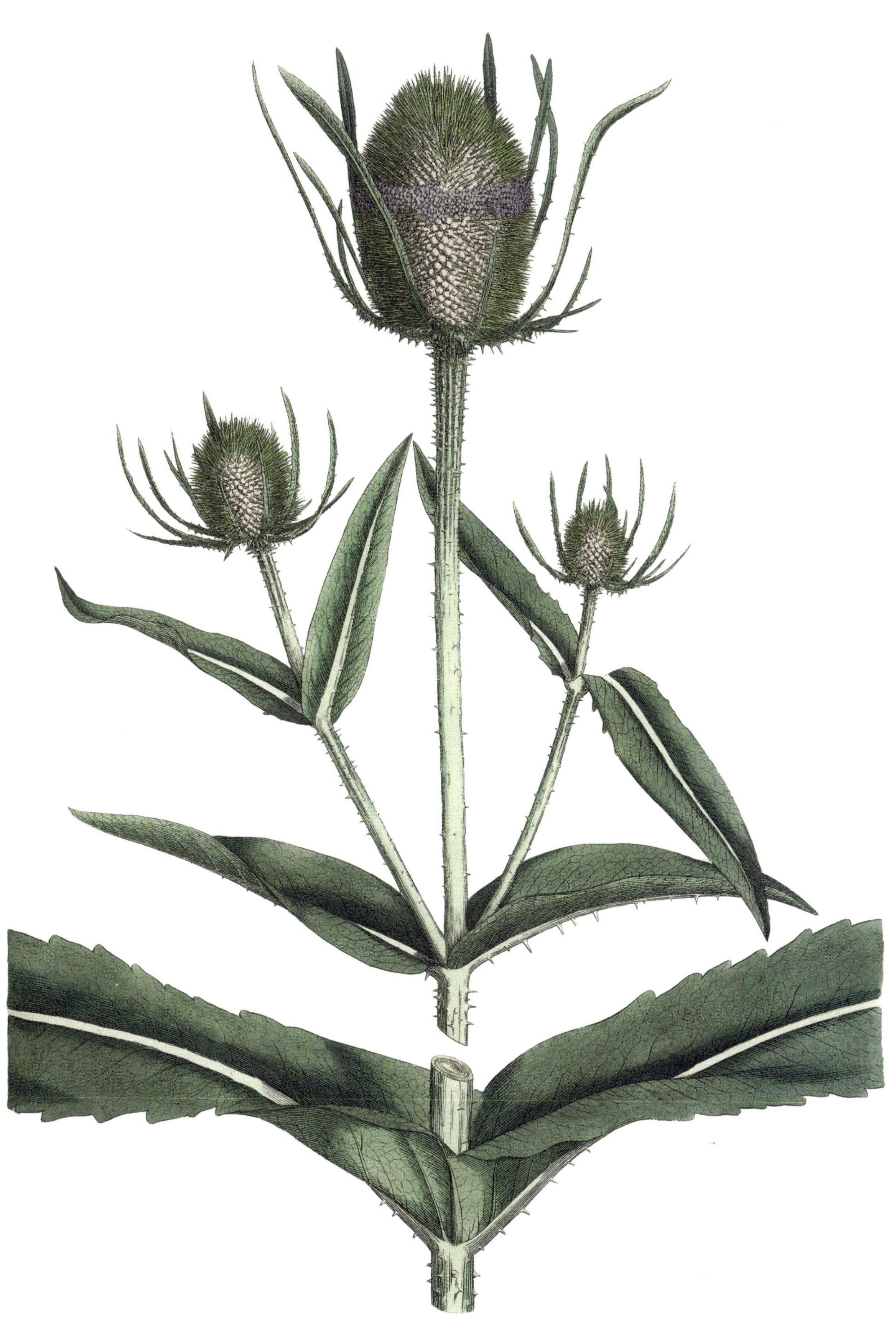

Dipsacus fullonum

ORPINE

Hylotelephium telephium

Like city workers hunting a patch of grass for some lunch-break sunshine, orpine is a sun-worshipper, seeding itself into the driest, sunniest spot it can find. You could almost fancy that its deep-pink flowerheads are due to sunburn. Thick, succulent leaves defend it from drying out, while its tubers store more fluid underground, meaning it can survive lengthy stretches without water. Tough and extremely low maintenance, it's perfect for the hot courtyard gardens common in cities.

ANATOMY NOTES

FLOWERS — Its flowers are often abuzz with a cacophony of bees, butterflies and other pollinators.

LEAVES — Orpine has adapted to arid conditions by photosynthesising during the night, rather than the day, helping it to retain moisture. Leaves can be crushed and applied to wounds and insect bites, but children prefer blowing them up like a bloated frog. Young leaves are edible.

ROOTS — You can boil and eat the roots when they're young in spring. Crush fresh roots to treat bruised skin, covering with a bandage for several hours.

STEMS — On midsummer's eve, young maidens would hang two stems of orpine in their doorway. If by morning the branches had bent towards each other, the prospective lover was true; if not, prospects weren't good.

BOTANIST'S TIP

You can grow wild orpine in the garden or try its cultivated cousin 'Purple Emperor', which has stunning purple leaves and stems. For a showy display, plant the cultivar 'Herbstfreude' (Autumn Joy) which has giant pink flowerheads. Orpine stems stay upright throughout autumn and winter and can be left in situ for architectural interest.

ESSENTIALS

Aka — Frog's-stomach, midsummer men

Family — Crassulaceae (stonecrop)

Urban habitat — Dry grassy areas and waste ground

Height — 60 cm / 2 ft

Flowers — July to August

Uses — Ornamental, food, medicinal

Grow — Sow seed in early spring

Harvest — Divide plants in spring

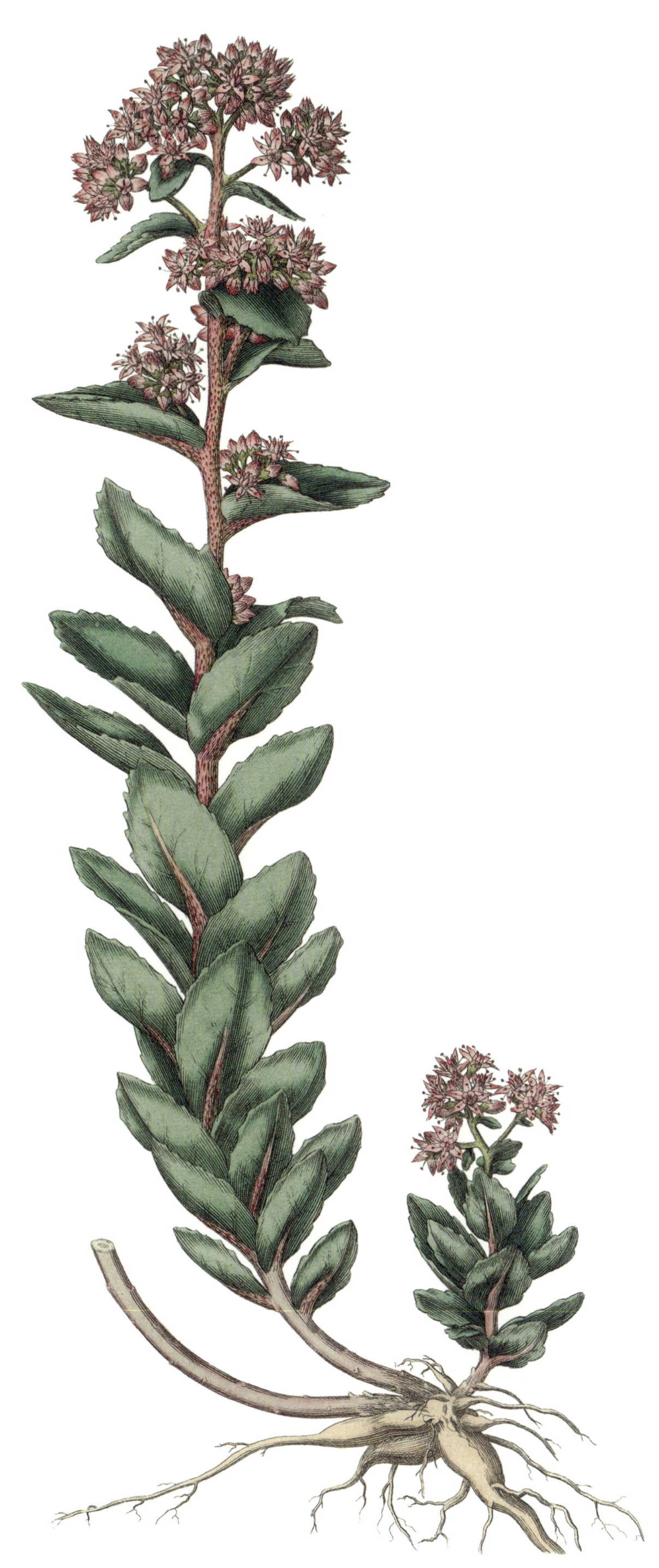

Hylotelephium telephium

SCARLET PIMPERNEL

Lysimachia arvensis

They seek him here, they seek him there; like the fictional Scarlet Pimpernel, the flower itself is sometimes 'demned elusive'. Fast disappearing from our cornfields, these days you might see its supremely delicate blooms pinpricking roadside verges crimson. Looking closely reveals a flower of great miniature beauty, splashed with magenta at the centre where stamens, crowned in yellow, are adorned with the finest hairs. When rain threatens, the petals close around this fragile arrangement for protection – forecasting bad weather ahead.

ANATOMY NOTES

FLOWERS — The flower is the famous pseudonym and symbol of the aristocratic dandy who leads a secret double life as the heroic daredevil in Baroness Emma Orczy's novel *The Scarlet Pimpernel*. In Irish folklore, the flower gives any bearer second sight and hearing, and the ability to understand birds and animals.

LEAVES — Herbalists used a tincture of this plant against melancholy and venomous bites. Modern research has discovered that the plant has good antimicrobial and anti-inflammatory properties. The saponin-laden leaves, a toxin, don't make for a decent foodstuff, even when cooked.

BOTANIST'S TIP

There's a bright blue kind of scarlet pimpernel, called Caerulea, that occasionally appears in the wild, usually in hot places. As the seeds are hard to come by, try the closely related 'Blue Cascade' for a similarly vivid blue.

ESSENTIALS

Aka — Poor man's weatherglass, shepherd's clock

Family — Primulaceae (primrose)

Urban habitat — Roadsides, verges, wasteland

Height — 40 cm / 1 ft 4 in

Flowers — July to August

Uses — Garden ornamental

Grow — Sow seed in spring

Few possess more liveliness of colour, or greater delicacy of structure

W. Curtis

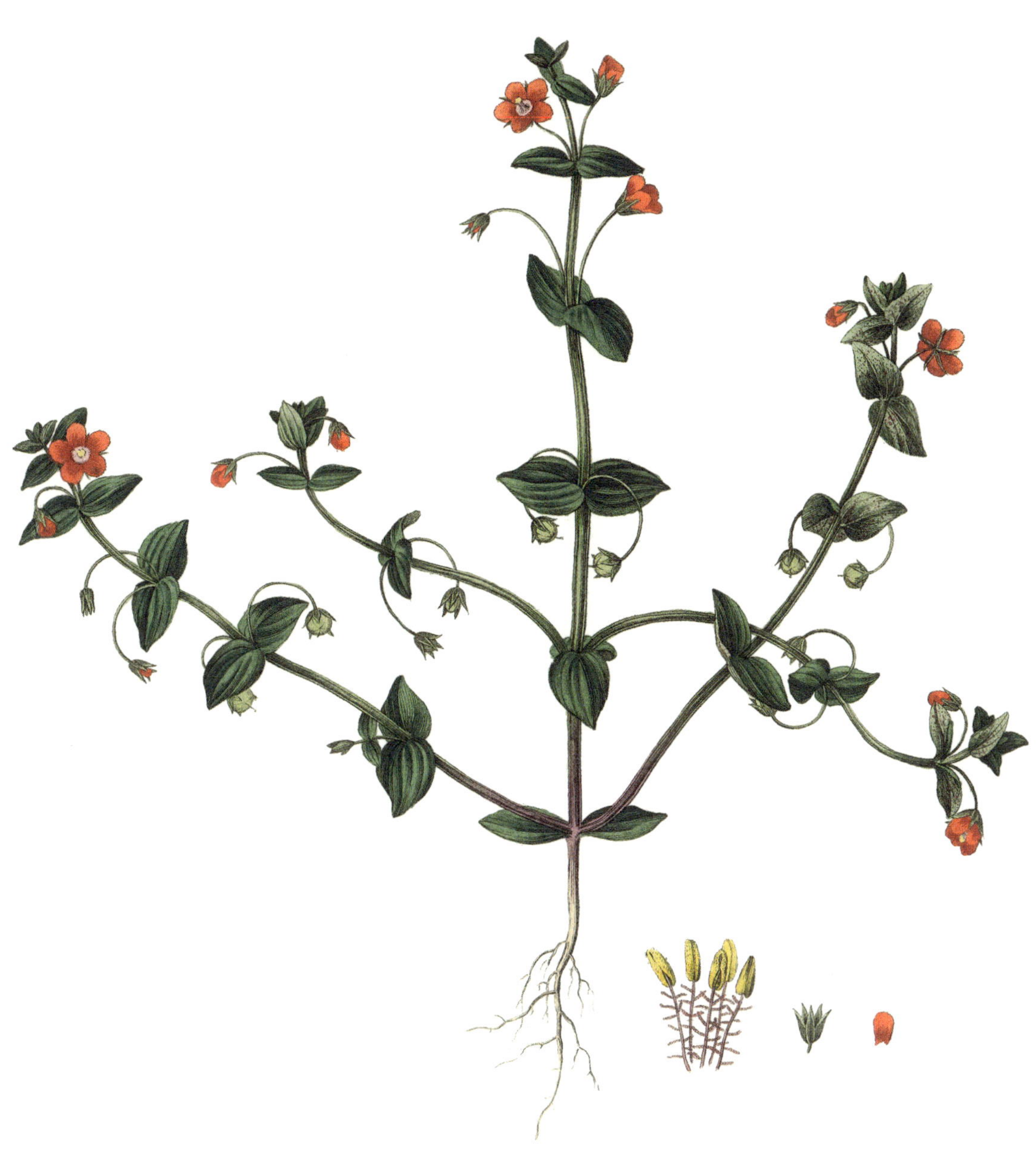

Lysimachia arvensis

FLOWERING RUSH

Butomus umbellatus

Tall, slender and topped in summer by graceful pink flowerheads, this aquatic plant lends a dash of sophistication to any waterside scene. A large cluster of them, handsomely swaying in the breeze, could make the boggiest ditch behind a supermarket car park look good. So, in a garden pond, and boosted by the shelter and heat of the typical small urban space, they'll be positively ravishing – and bring bees and dragonflies into the bargain. But it's a plant with an edge – literally. Its triangular leaves are sharp enough to deter cows from munching it.

ANATOMY NOTES

FLOWERS — The flowers are fragrant and were once used as a strewing herb to perfume the home.

LEAVES — When the leaves are too long to support their own weight, they often fall upon the bank and provide an improvised ladder for damsel and dragonfly nymphs. The sabre-edged leaves can cut the tongues of any cattle looking for a nibble. That's how the plant gets its scientific name *Butomus*, which literally means 'cow cutter'.

ROOTS — As they're more than 50% starch, you could dry the roots and grind them into flour or roast them like a potato – if you really had to.

BOTANIST'S TIP

The rhizomes (underwater shoots) of the flowering rush will gradually multiply. If they start to take over a small pond, in spring you should lift the rhizomes out of the water before they start to grow leaves and stems, and snap them to divide the plants. Place half of the plants back in the water and pot up the rest into aquatic compost to give away.

ESSENTIALS

Aka — Water gladiolus

Family — Butomaceae (flowering rush)

Urban habitat — Wetlands and water margins

Height — 1.2 m / 3 ft 11 in

Flowers — July to September

Uses — Ornamental pond plant, edible root

Grow — Sow seed in autumn, or buy a plant and sink it into your pond before spring

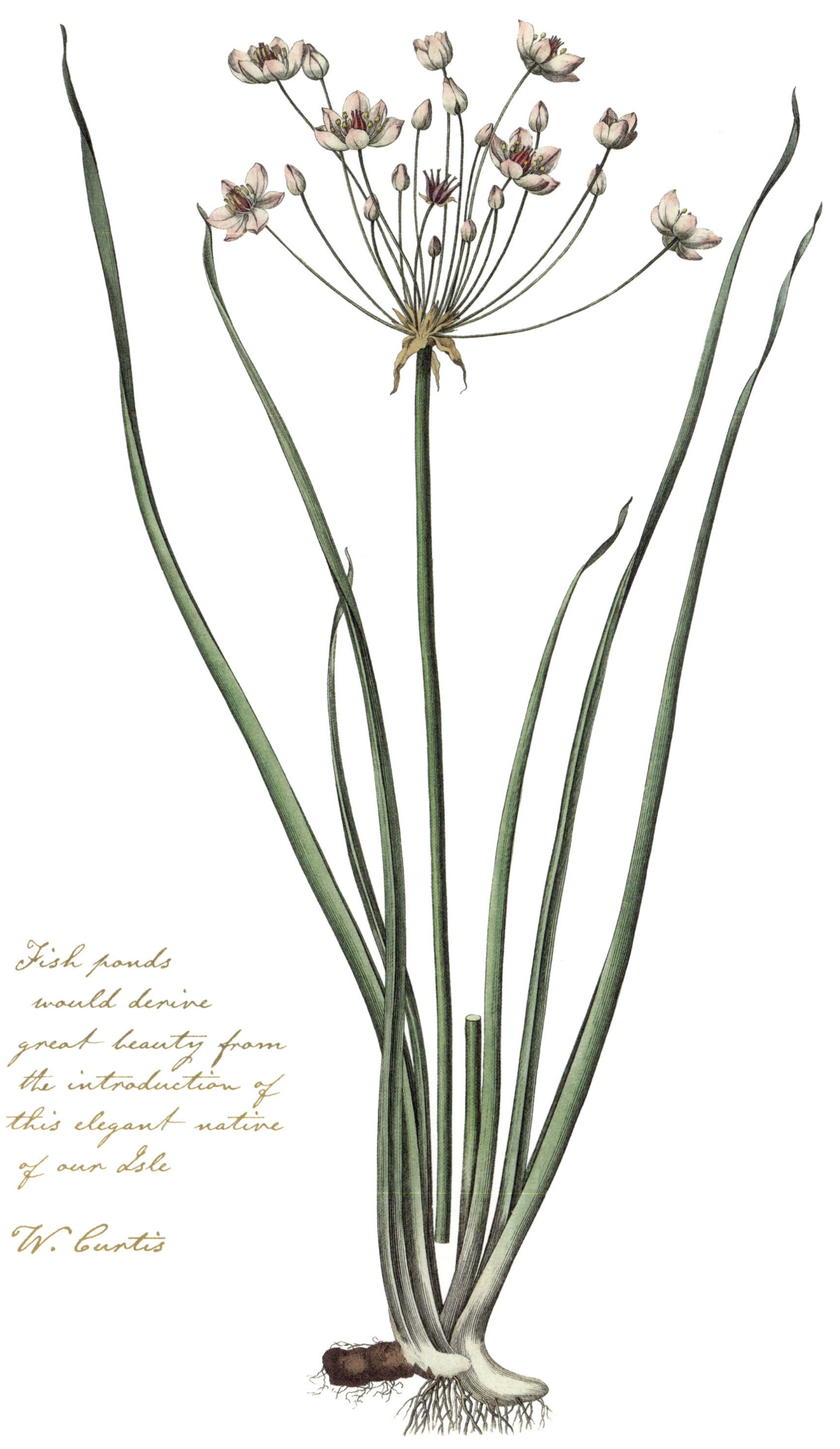

Butomus umbellatus

IVY

Hedera helix

Long before the present vogue for vertical gardens, ivy was scaling high walls, clothing any structure it could with its dark glossy leaves. On buildings it not only adds a touch of evergreen class, but also improves insulation by more than a third, and protects against damp and pollutants. But it's also a ferocious climber and if neglected can loosen poor brickwork and choke gutters with thick woody stems. Contrary to popular belief, it's not a parasite and doesn't strangle trees – but it does provide food and shelter for all kinds of insects, bird, bats and mice. And it has a few little-known medicinal uses too, including a 9th-century remedy for sunburn.

ANATOMY NOTES

FLOWERS — Mature plants flower late in the year, just in time for insects to stock up on food before hibernating.

FRUITS — Black berries appear around Christmas and ripen through winter, providing much needed calories and nutrients for birds.

LEAVES — Young leaves have the classic ivy shape, but mature ones are oval or heart-shaped. The leaves contain saponins, a soapy chemical; boiling them with a spoonful of sodium carbonate for 5 minutes makes a laundry liquid. Anglo-Saxons boiled ivy leaves in butter to relieve sunburn, and the Romans believed ivy wreaths stopped you getting drunk. Bacchus, the god of wine, always sports one, yet appears relentlessly 'cheerful'.

ROOTS — As well as sturdy ground roots, ivy's aerial ones secrete tiny particles which effectively glue it to surfaces. These particles absorb UV light and have potential to be used in sunscreens.

BOTANIST'S TIP

Ivy is nature's makeover artist, transforming dry ground and ugly walls into something of stately luxuriance. When training it up a wall, use a trellis set away from the building to give the aerial roots something else to cling to. This also maximises its thermal properties, warming walls in winter and cooling them in summer. There are some beautiful ivy cultivars to try, such as the yellow-leaved 'Buttercup' or the variegated (white-edged) 'Little Diamond'.

ESSENTIALS

Aka — English ivy

Family — Araliaceae (ivy)

Urban habitat — Walls, lampposts, buildings, shady parks and gardens

Height — 12 m / 40 ft

Flowers — October to November

Uses — After sun, cough medicines

Grow — Buy a plant in autumn

Harvest — Young leaves and twigs in spring

Trees covered with ivy have a very pleasing effect, and moreover induce birds of song to haunt those thickets

W. Curtis

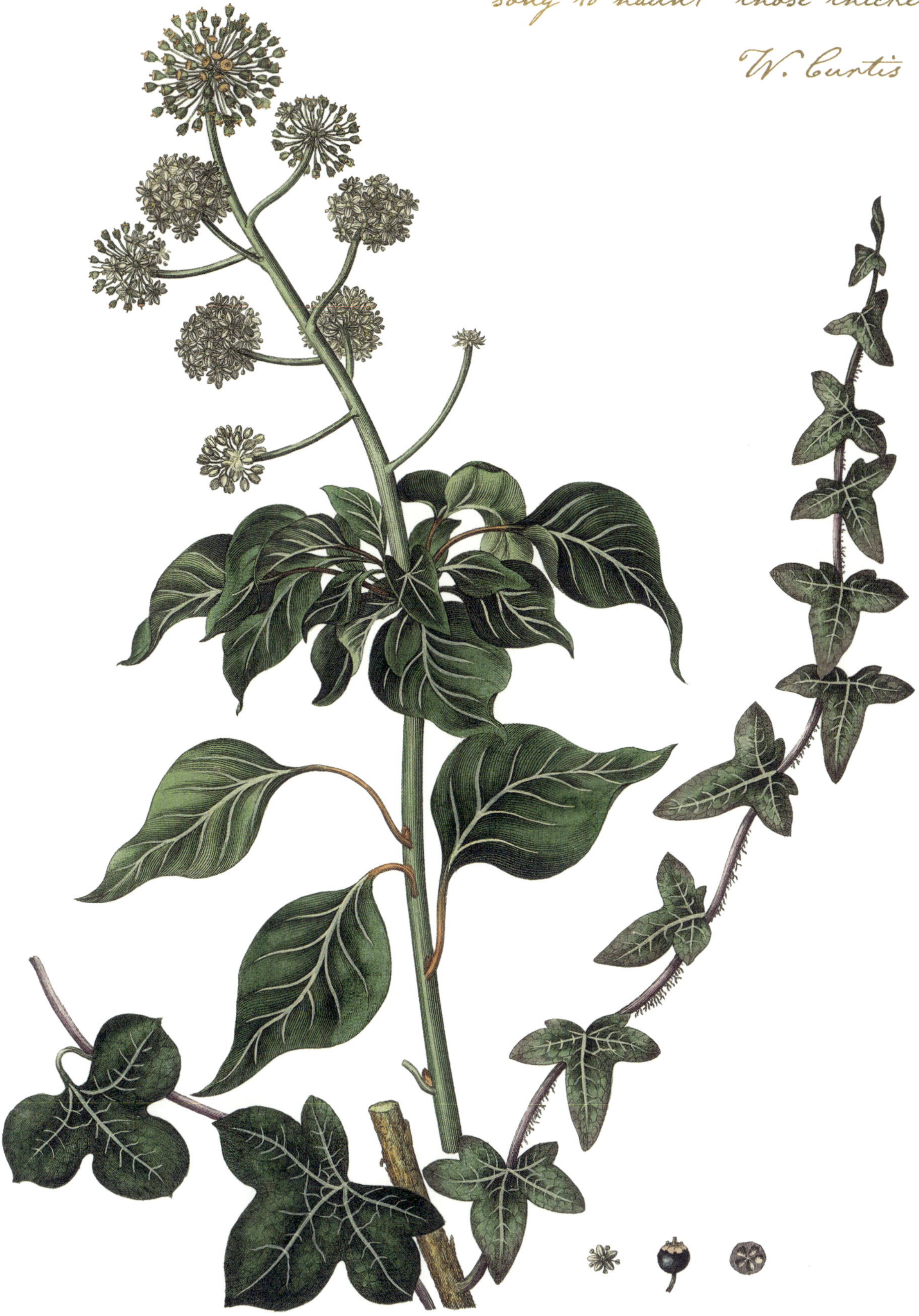

Hedera helix

SWAN'S-NECK THYME-MOSS

Mnium hornum

The deep-green velvet of swan's-neck thyme-moss smooths and softens the hard edges of canal bridges and damp brick walls all over town. It's a common sight, though mosses are tricky to identify from a distance. Get close and you'll see that its leaves resemble shrunken ferns, and closer still, you'll spot crowds of tiny stalks (seta) topped by drooping spore capsules that make each bend like a swan's neck. New leaf growth starts out acid green, darkening as it matures. In any garden, it provides texture and softness, but you could go the whole hog and create an entire Japanese moss garden, famed for their lush tranquillity – and easy upkeep.

ANATOMY NOTES

CAPSULE — A lid on the capsule falls off when the spores are ready. When weather is dry, tiny rows of teeth open, allowing the wind to disperse them.

LEAVES — The dense leaves form a miniature ecosystem for scores of insects, invertebrates, and their predators. They also make a cosy insulator to line the homes of larger animals, like mice and squirrels; even we humans once used it to stuff mattresses. The leaves' fatty acids are beneficial for the immune system. Moss needs very damp conditions to survive as it has no way of retaining water. Like a kind of sponge, the leaves constantly soak up water through tiny pores.

RHIZOIDS — Technically, mosses don't have roots but rhizoids, which anchor the plant to its chosen surface. They don't take up water, as mosses don't have a vascular system.

BOTANIST'S TIP

This is a great plant for a base layer in a terrarium, a glass container filled with plants to make a mini tabletop garden. Team with ferns, ivies and other moisture-loving plants that thrive in closed environments. Get plants online to avoid bringing in unwanted insects from the outside.

ESSENTIALS

Family — Mniaceae (thyme-moss)

Urban habitat — Damp, shady brickwork and tree stumps

Height — 5 cm / 2 in

Flowers — No flowers, but green all year

Uses — Gardens, terrariums, mattress stuffing

Grow — Plant moss sections in spring

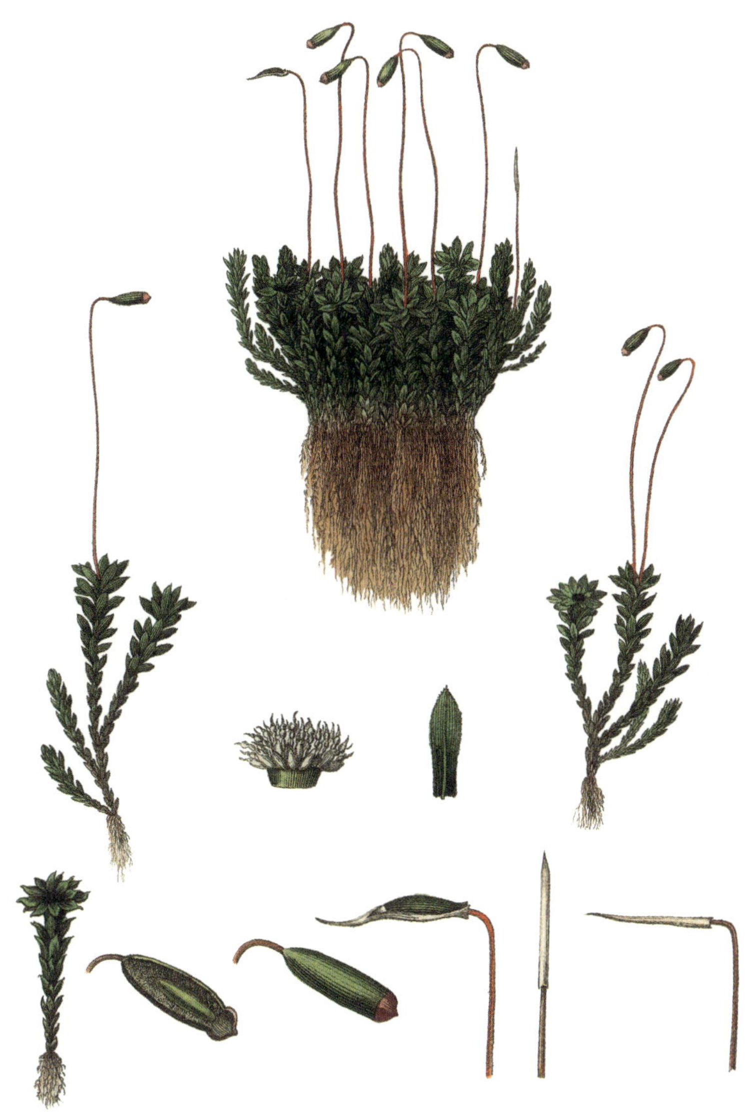

Mnium hornum

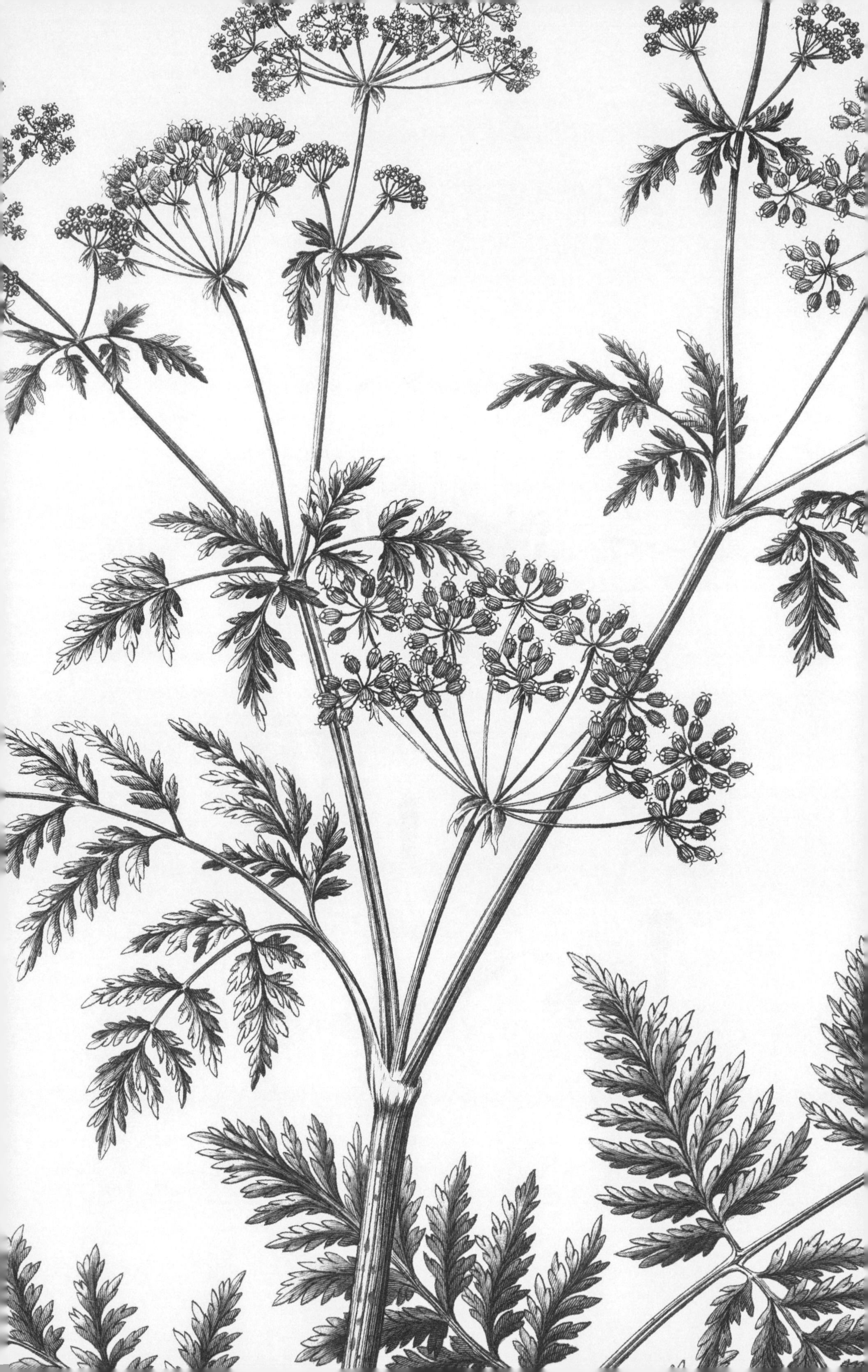

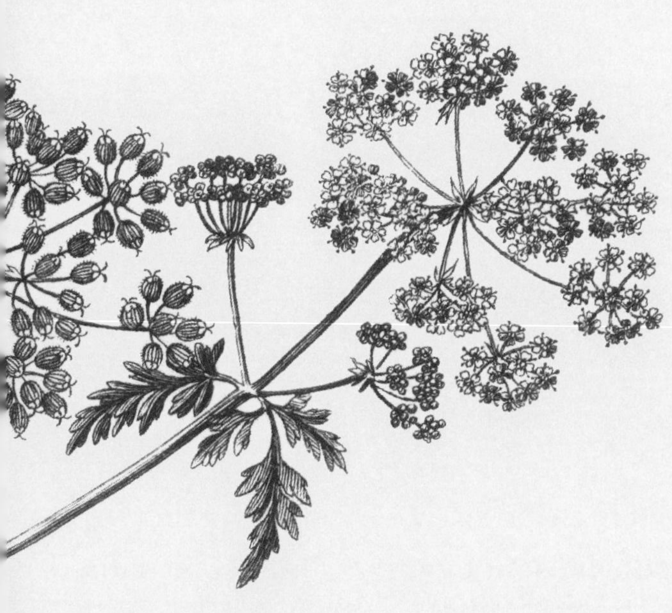

KILL

BEWARE the dark side.

These floral *FEMMES FATALES*

are among the most beautiful plants of all

and the *DEADLIEST*.

Don't say you weren't *WARNED*.

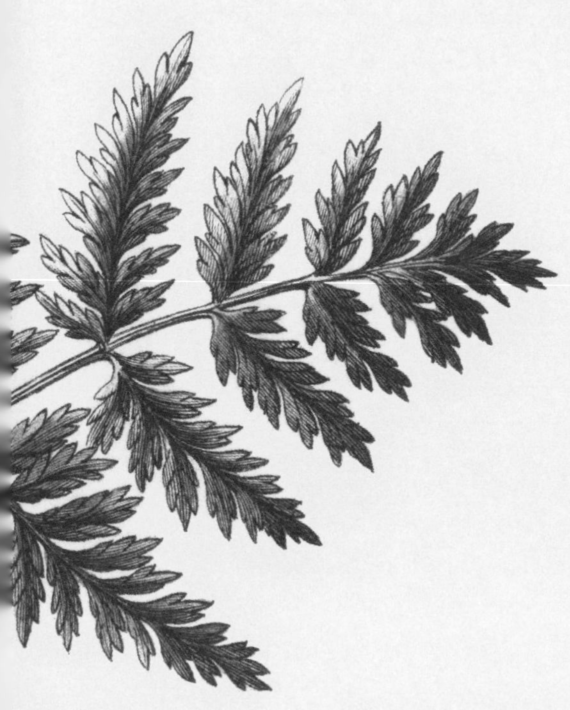

HEMLOCK

Conium maculatum

If there's any plant that brings foragers out in a cold sweat – even before they've accidentally eaten it – it's hemlock. Deadly poisonous and easily mistaken for other plants in the carrot family that make a good meal, it's the bogeyman that any gatherer must learn to identify. Left to its own devices – as it generally should be – hemlock can grow in large thickets and raise white blooms to well over 2 m / 6 ft, the UK's tallest native umbellifer. Lofty, imposing, and once the executioners' poison of choice, hemlock always demands respect.

ANATOMY NOTES

FLOWERS — The flowers are a great source of nectar for many insects, so it is a useful plant – not something to fear and destroy.

LEAVES — Crushed leaves give off an unpleasant musty smell, which should warn most sensible people that this isn't a plant to eat. Until the beginning of the First World War, Britain exported tonnes of dried leaves and seeds of hemlock to the US for use against asthma, epilepsy and whooping cough, illnesses that are no longer treated in this way.

STEM — All mammals, not just humans, are poisoned by hemlock. Grazing animals even steer clear of dead stems, which remain toxic for several years. This plant really doesn't want to be munched. When mature, the live stem has purple blotches down it, a distinguishing feature; *maculatum* in Latin means spotted.

ESSENTIALS

Aka — Poison parsley

Family — Apiaceae (carrot)

Urban habitat — Sunny grassy places, roadsides, watersides

Height — 2.5 m / 8 ft 2 in

Flowers — June to July

Uses — Poison

BOTANIST'S WARNING

Hemlock contains coniine which causes muscular paralysis and leads to a death by asphyxiation as the muscles controlling the respiratory system fail. The plant was used by the Ancient Greeks for executing prisoners – most famously the philosopher Socrates, who was convicted of corrupting young minds. Plato recorded how the poison progressively numbed his body from his feet upwards until it reached his heart, killing him.

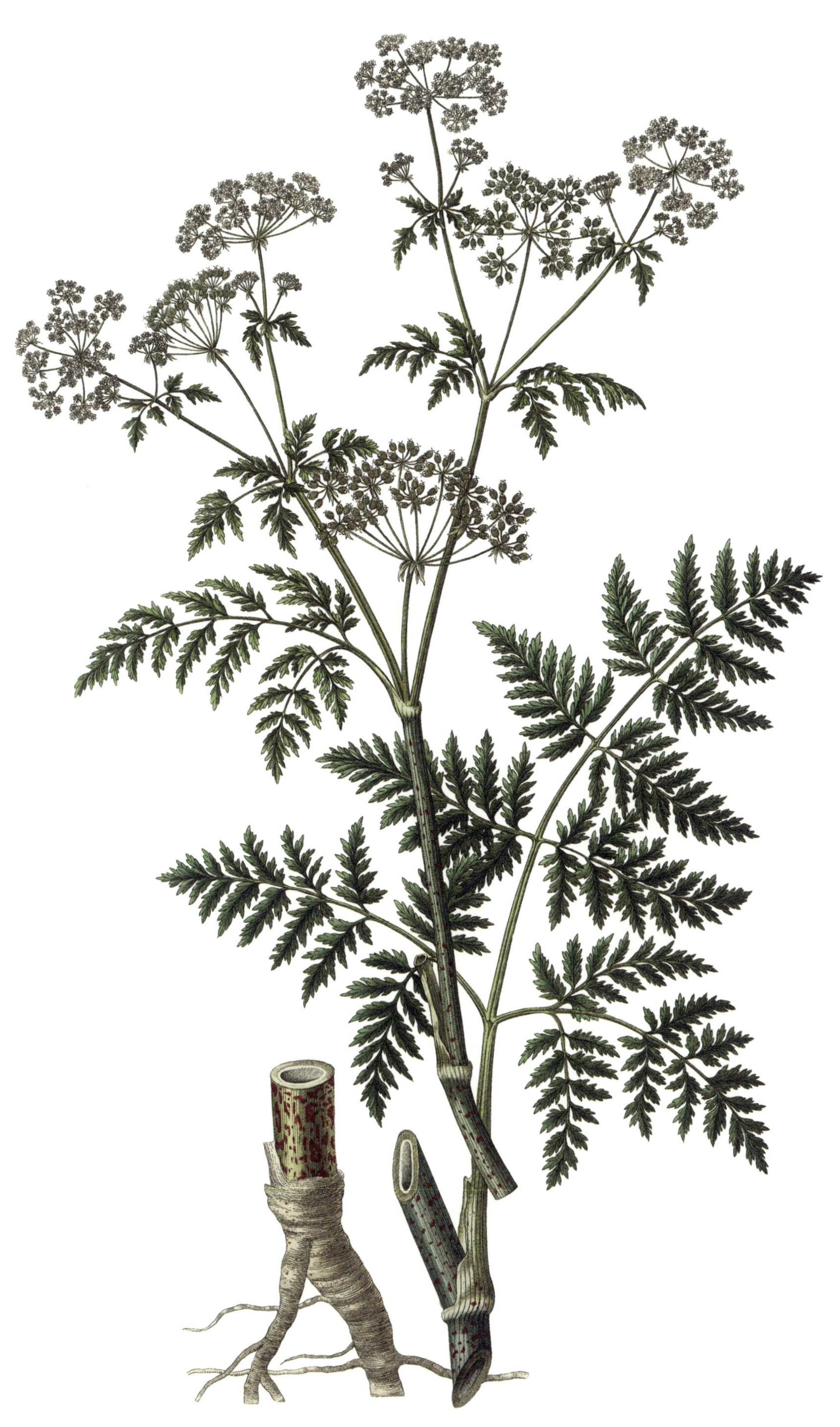

Conium maculatum

LILY OF THE VALLEY

Convallaria majalis

Lily of the valley, the daintiest of blooms with the divinest scent, is a favourite wildflower of many people – not just perfumiers and florists. Spotting its delicately dangling white bells in the shelter of a hedge is like finding strings of pearls scattered in the rough. Look, but don't touch. Its lethal poisons can bring you out in a nasty rash – and much worse if you accidentally eat it. Leave that to the bees who also love its fragrance, and the birds that gorge on its red berries in autumn.

ANATOMY NOTES

FRUITS — Highly toxic to mammals, the berries are nevertheless enjoyed by birds, which eat, carry and pass out the seeds elsewhere.

FLOWERS — Symbolising love, luck and happiness, its delicate flowers commonly feature in springtime bridal bouquets. They don't produce much essential oil, so perfumiers often imitate its heavenly fragrance with synthetic chemicals. Christian Dior adored the flower, so it's not surprising that his 1956 perfume, *Diorissimo*, is still regarded as one of the best.

LEAVES — The leaves can be used to produce textile dyes, giving a light green colour when harvested in spring and yellow when collected in autumn. In traditional herbal medicine, the leaves were used to create a tonic to counter mild heart complaints – but the safe procedure demanded expert understanding of preparation and dosage.

BOTANIST'S WARNING

Lily of the valley contains cardenolide, a steroid which in high concentrations causes the heart to stop. The most likely reason anybody might ingest lily of the valley is through misidentification. It often grows freely alongside wild garlic (*Allium ursinum*), a commonly foraged plant with very similar leaves.

ESSENTIALS

Aka — Our lady's tears

Family — Asparagaceae (asparagus)

Urban habitat — Under dry hedges, dappled woods

Height — 25 cm / 10 in

Flowers — May to June

Uses — Perfume

Grow — Buy for planting in autumn

Harvest — Flowers in late spring; wear gloves

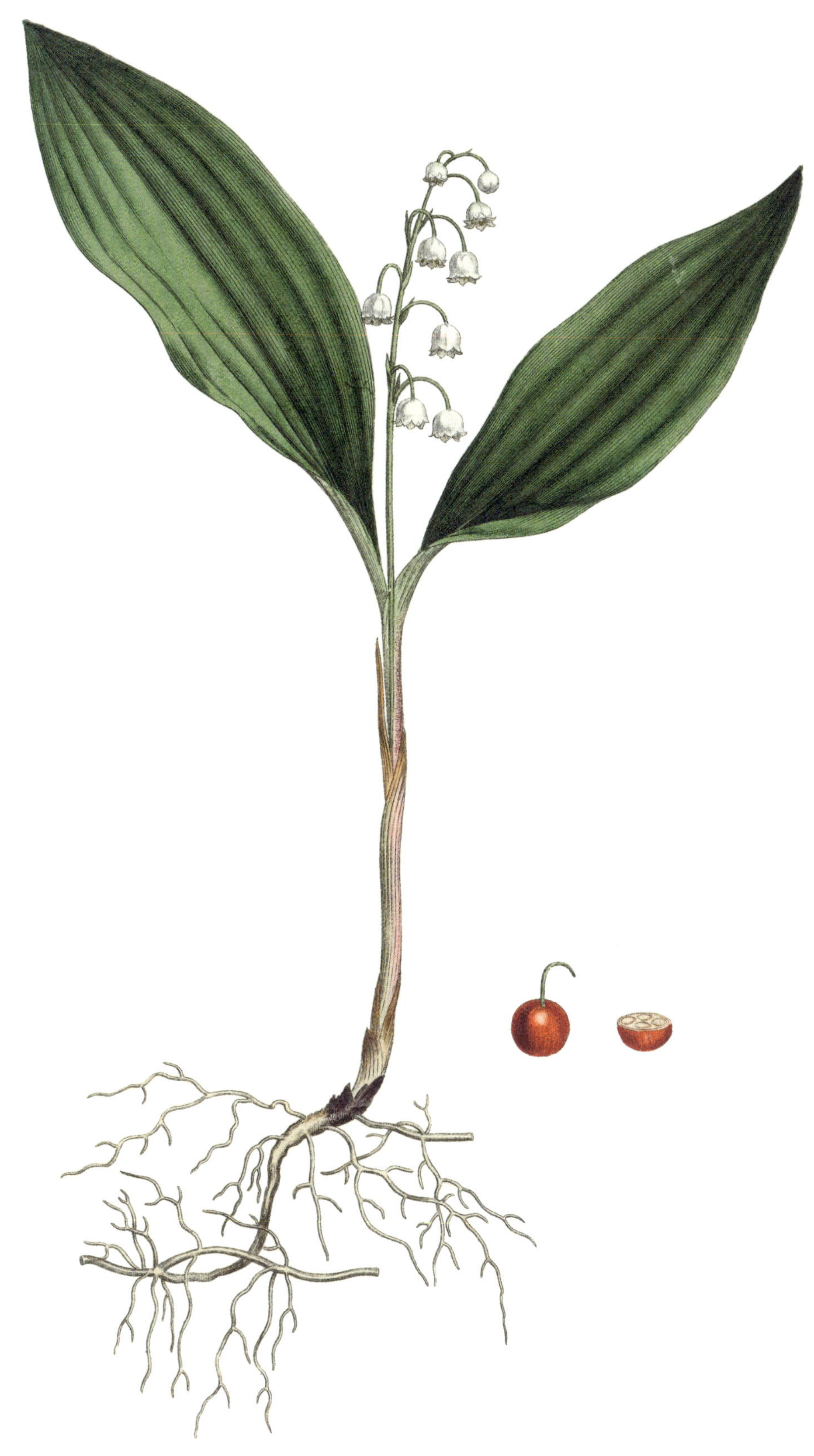

Convallaria majalis

BULBOUS BUTTERCUP

Ranunculus bulbosus

These sunny yellow buttercups, once common on pasture and hay meadows, and now also spotted on roadside verges in the city, have long been the bane of farmers. Grazing animals that haven't yet learnt it soon discover that this is the most poisonous buttercup of all. Not a plant to pick for posies, it will blister your skin – a property herbalists once made use of to treat stiff joints, and beggars to provoke pity. It grows from a bulb-like tuber, called a corm, which only pigs enjoy eating.

ANATOMY NOTES

FLOWERS — Just below the surface of the petal there is a layer of air which gives the flower an extreme glossiness. This acts almost as a mirror and serves to attract insects.

ROOTS — The roots are edible cooked but not very pleasant, so have only been harvested as a famine crop. The rashes and blistering caused by eating the root raw reminded people of St Anthony's fire, which could refer to the skin infection, erysipelas. Believers prayed to the saint for a cure – hence the plant's other common name of St Anthony's turnip. The raw root was once chewed up and put into tooth cavities to relieve toothache; maybe the pain of the blistering just distracted the patient.

LEAVES — The leaves were applied by herbalists to treat some joint and skin conditions, including arthritis, eczema and gout. There are now better ways of managing these ailments without the risk of a burn.

ESSENTIALS

Aka — St Anthony's turnip

Family — Ranunculaceae (buttercup)

Urban habitat — Grassy areas and verges

Height — 30 cm / 1 ft

Flowers — March to June

BOTANIST'S WARNING

The bulbous buttercup contains ranunculin, as do many plants in the buttercup family, which turns into the toxin protoanemonin when the plant is damaged. This causes blistering of the skin or, if eaten, nausea, muscle spasms and worse.

Hogs are fond of the roots and will frequently dig them up

W. Curtis

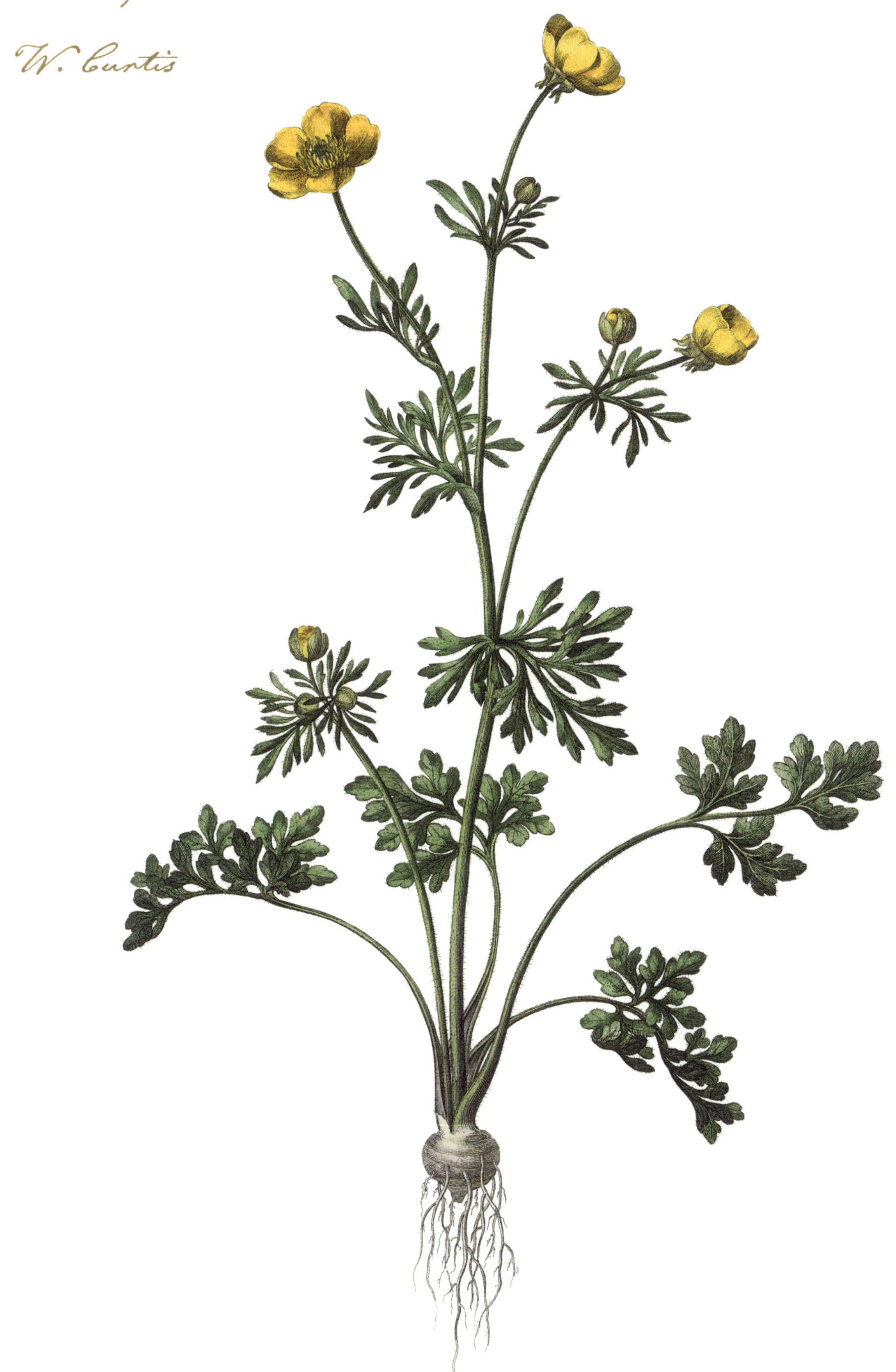

Ranunculus bulbosus

DEADLY NIGHTSHADE

Atropa bella-donna

Deadly nightshade is the most infamous plant of them all, the poison of choice for history's villains and the killer of kings and emperors. A true *bella donna*, or beautiful lady, clothed in lush foliage and voluptuous purple flowers, she tempts the unwary with shiny and sweetly seductive berries. It's a fruit you'd only eat once, though: her cloying aftertaste is followed by blurred vision, slurred speech, delirium, convulsions and ultimately, death. The warnings are clear in her name, common and scientific; *Atropa* is from Atropos, the Greek Fate who cuts the thread that ends a mortal's life. Don't let it be yours.

ANATOMY NOTES

BERRIES — Women of yore squeezed the berry juice into their eyes to make their pupils dilate attractively, a possible origin of the *bella-donna* part of the Latin name. Although not used cosmetically, its extracts are still sometimes used by surgeons during eye operations. The berries look a little like cherries, but are definitely not right for your black forest gateau. In folklore, deadly nightshade is the devil's own plant. So, if you eat his berries, in return you can expect the severest punishment there is.

LEAVES — In the middle ages, the leaves along with the berries were used to create a blush for ladies' cheeks – another reason that the plant is associated with feminine beauty.

ROOTS — Extracts from the roots are used in modern medicine to relieve muscle spasms.

BOTANIST'S WARNING

Deadly nightshade contains tropane alkaloids which can wreak havoc on the human body. Medical students learn some of the symptoms of this type of a poisoning with a mnemonic: 'Blind as a bat, dry as a bone, hot as a hare, red as a beet, mad as a hatter...' After the 'mad' hallucinations and delirium, there might be coma, and if there's a recovery, it often comes with terrible paranoia until the chemical wears off.

ESSENTIALS

Aka — Devil's berries, naughty man's cherries

Family — Solanaceae (nightshade)

Urban habitat — Dappled shade

Height — 2 m / 6 ft 7 in

Flowers — June to August

Uses — Commercial medicines

Grow — Sow seed in spring

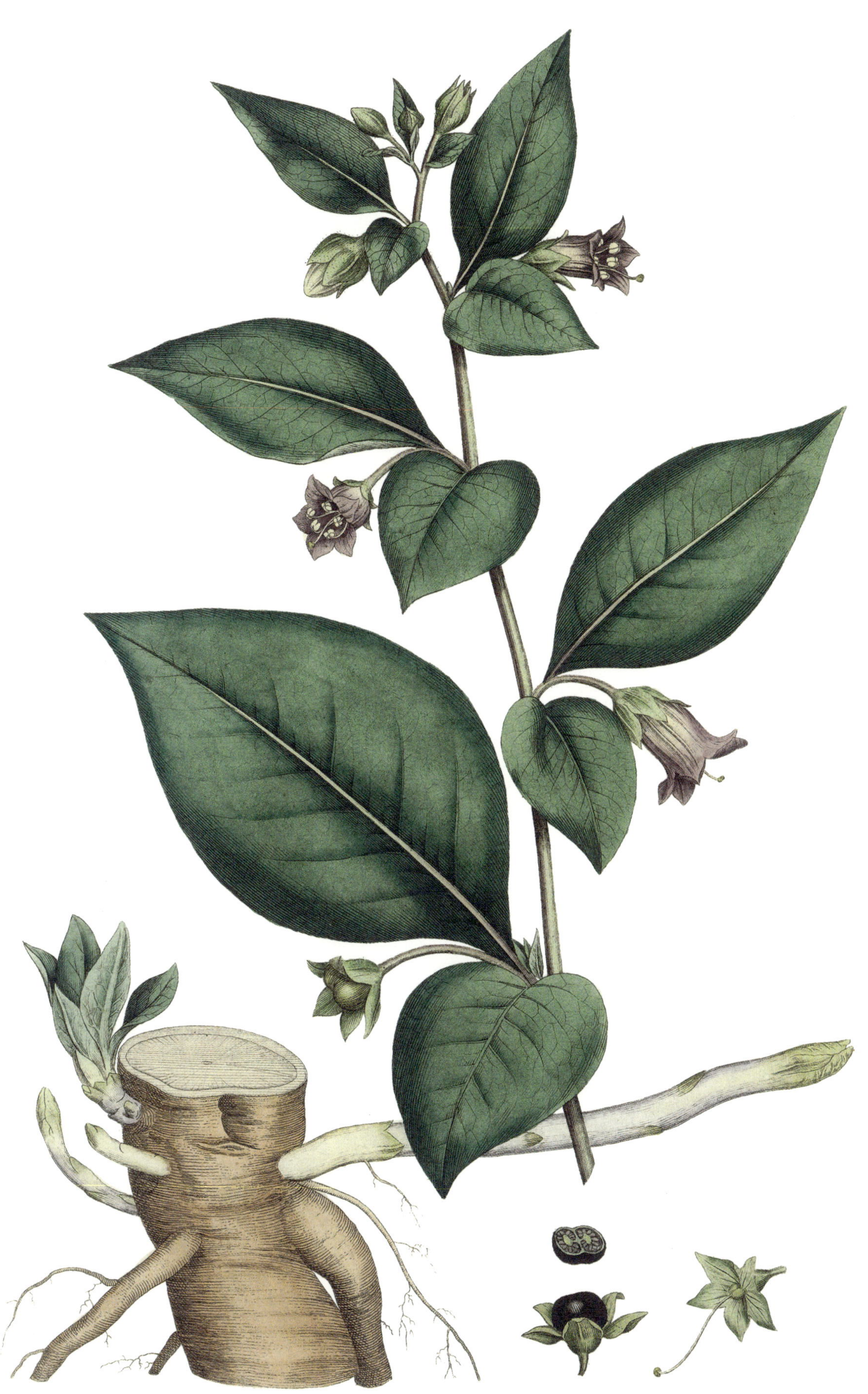

Atropa bella-donna

PHEASANT'S EYE

Adonis annua

Crimson blooms with unblinking black centres, as startling as a beady bird's eye, used to pepper our cornfields with colour. Since the Iron Age, this pheasant's eye was successful at slipping its seeds in with the farmer's crop – and, come summer, young entrepreneurs would gather the bright little flowers and feathery foliage for sale in town markets. Modern herbicides have all but done for poor pheasant's eye – as it will do for you if you eat this poisonous beauty. It now clings on in rare patches, a scrubby road verge or neglected roundabout, where soil containing old seeds gets disturbed.

ANATOMY NOTES

SEEDS — The seeds are spread by ants, helping wide dispersal, and can lie dormant in the soil for many years, waiting for the ground to be disturbed.

FLOWERS — The plant is named after Adonis, a character in Greek myth so handsome that the goddess Aphrodite herself fell in love with him. He was killed by a wild boar and red flowers sprang from the ground where his blood fell. They're usually taken to be anemones, which bear more than a passing resemblance to their cousin, blood-red pheasant's eye. It's called *annua* (annual) because the plant completes its lifecycle inside a year.

LEAVES — Patients who took pheasant's eye as a traditional cure for sexually transmitted diseases often risked an outcome that was worse than their original affliction – death.

BOTANIST'S WARNING

Pheasant's eye contains cardiac glycosides which interfere with heart rhythm and can stop it beating altogether with a high enough dose. When extracted under controlled methods and administered by an expert, the chemical can actually help cure mild heart conditions, but the margin of error is a fine one.

ESSENTIALS

Aka — Red Morocco

Family — Ranunculaceae (buttercup)

Urban habitat — Disturbed ground, verges, roadsides

Height — 40 cm / 1 ft 4 in

Flowers — June to July

Uses — Cut flower

Grow — Sow in autumn

Harvest — Flowers in summer

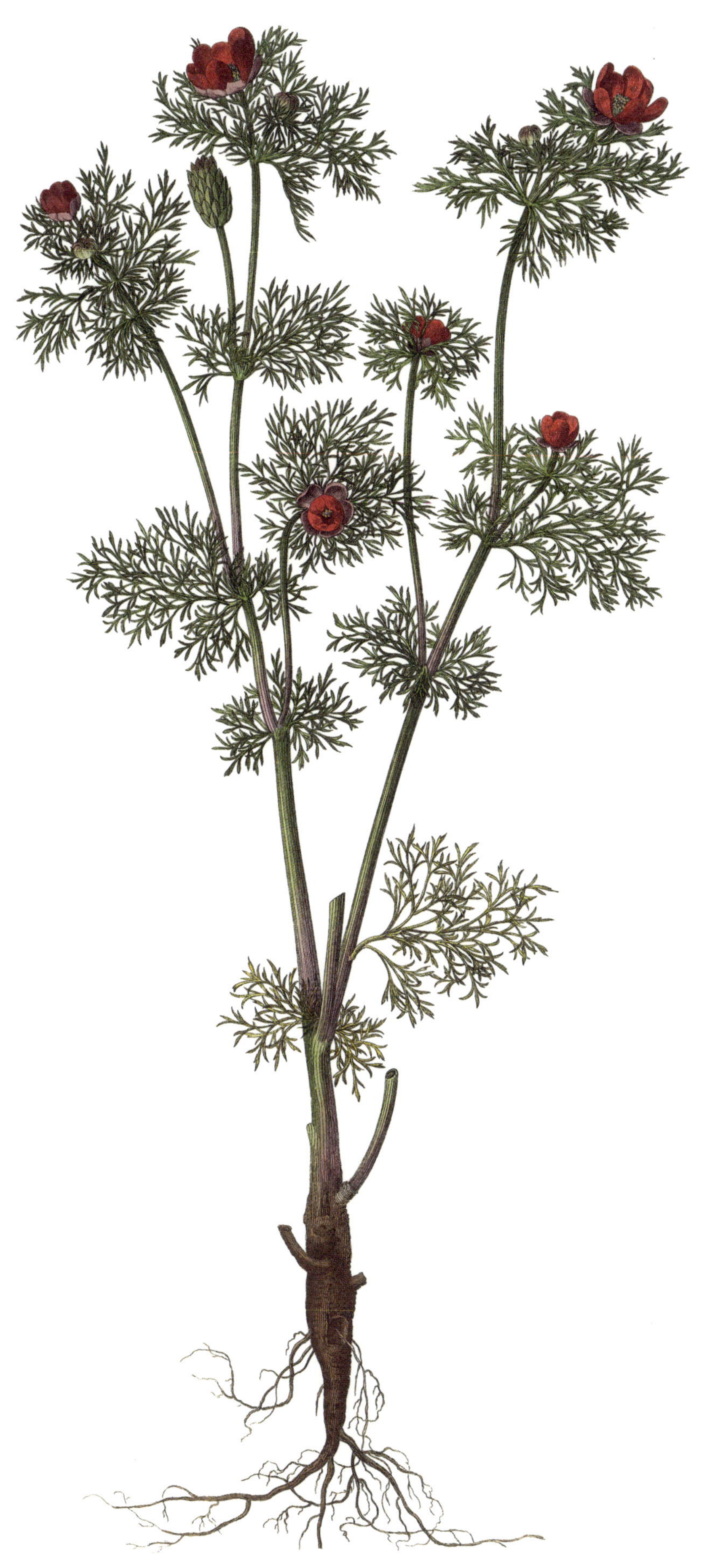

Adonis annua

THORN APPLE

Datura stramonium

The thorn apple's outsized trumpet flowers, broad prickly leaves and ferociously spiky seed capsules make for a distinctly tropical get-up – just right for a highly dangerous plant that probably originally hails from the jungles of Central America. A UK resident for at least 350 years, it's now found on waste ground all over town. But this is an apple that's definitely not for eating. Those unfortunate enough to ingest any part of the plant can expect to find themselves gibbering, confused and hallucinating for days – if they survive at all.

ANATOMY NOTES

SEEDS — The seeds have the strongest concentration of toxins. Some cultures take them to induce visions and to commune with the dead. Individual plants vary widely in toxicity and it's easy to overdose if you don't know what you're doing. Too much seed and you could be joining the departed.

FLOWERS — The flowers are pollinated by moths, particularly those with long tongues that can reach the bottom of the trumpets. The Hindu god Shiva as Nataraja, the lord of the dance, wears a datura flower in his hair.

LEAVES — In 1676, British soldiers stationed at Jamestown, Virginia, ate a salad of thorn apple leaves and underwent 11 days of bizarre and delirious behaviour, including sitting naked and grimacing like a monkey. None remembered a thing when it was over. Because of this event, the plant is also known as jimsonweed (Jamestown weed) in the US.

ESSENTIALS

Aka — Jimsonweed

Family — Solanaceae (nightshade)

Urban habitat — Scrubland

Height — 70 cm / 2 ft 4 in

Flowers — July to October

Grow — Sow seed in autumn

BOTANIST'S WARNING

Thorn apples contain tropane alkaloids, chemical compounds with a range of important applications in pharmacology. But in the wilds and in the wrong doses, they can have devastating effects and can be fatal.

The symptoms are light-headedness, profound sleep, insanity, madness, convulsions, palsy of the limbs, cold sweats, vehement thirst and tremblings

W. Curtis

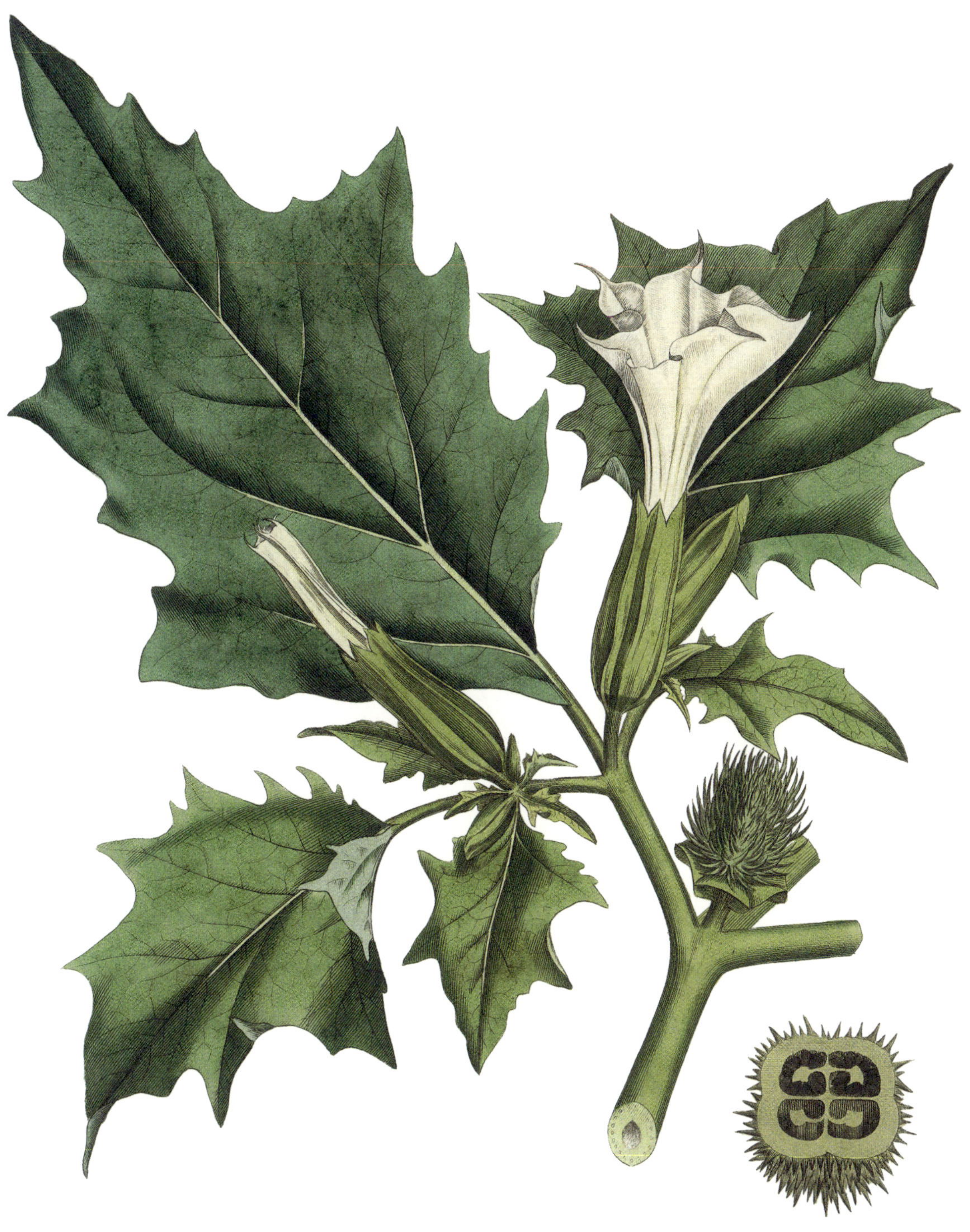

Datura stramonium

GREEN HELLEBORE

Helleborus viridis

A keen eye helps when searching for the elegant green flowers of this hellebore. Among the earliest blooms of the year, it stands modestly in the shade, camouflaged enough to reward only the most observant. It makes a gorgeous winter posy alongside their seasonal companions, snowdrops and crocus, but take care when handling – it's poisonous and can burn the skin. Bumblebees appreciate green hellebore as an early source of nectar, hunting it out even among the more colourful flowers of other early bloomers.

ANATOMY NOTES

FLOWERS — To attract pollinators in from the winter cold, green hellebores offers them warm alcoholic drinks in a fully heated flower – literally. A fermenting yeast in the nectar not only makes it mildly alcoholic but raises the flower temperature by as much as 6 °C / 43 °F. Hot toddies for bees in the green hellebar!

LEAVES — As the plant irritates mucus membranes, herbalists once dried and powdered the leaves to make a medicinal sneezing powder. It'll make you vomit if you eat it, so it became a traditional remedy for worms, a 'cure' that probably did more damage than the worms themselves.

ROOTS — In the past, the roots and rhizomes have been used as a laxative and as a solution against body lice, treatments long superseded by less toxic remedies.

BOTANIST'S WARNING

Green hellebore contains two toxic compounds: ranunculin, which is prevalent in the buttercup family and causes skin irritation; and bufadienolide, a steroid that raises blood pressure, leading in the worst cases to coma and cardiac arrest. A plant for gardeners rather than gourmets.

ESSENTIALS

Aka — Bastard hellebore

Family — Ranunculaceae (buttercup)

Urban habitat — Dappled shade in verges and woods

Height — 50 cm / 1 ft 8 in

Flowers — February to April

Uses — Cut flower

Grow — Buy for autumn planting

Harvest — Flowers in late winter; wear gloves

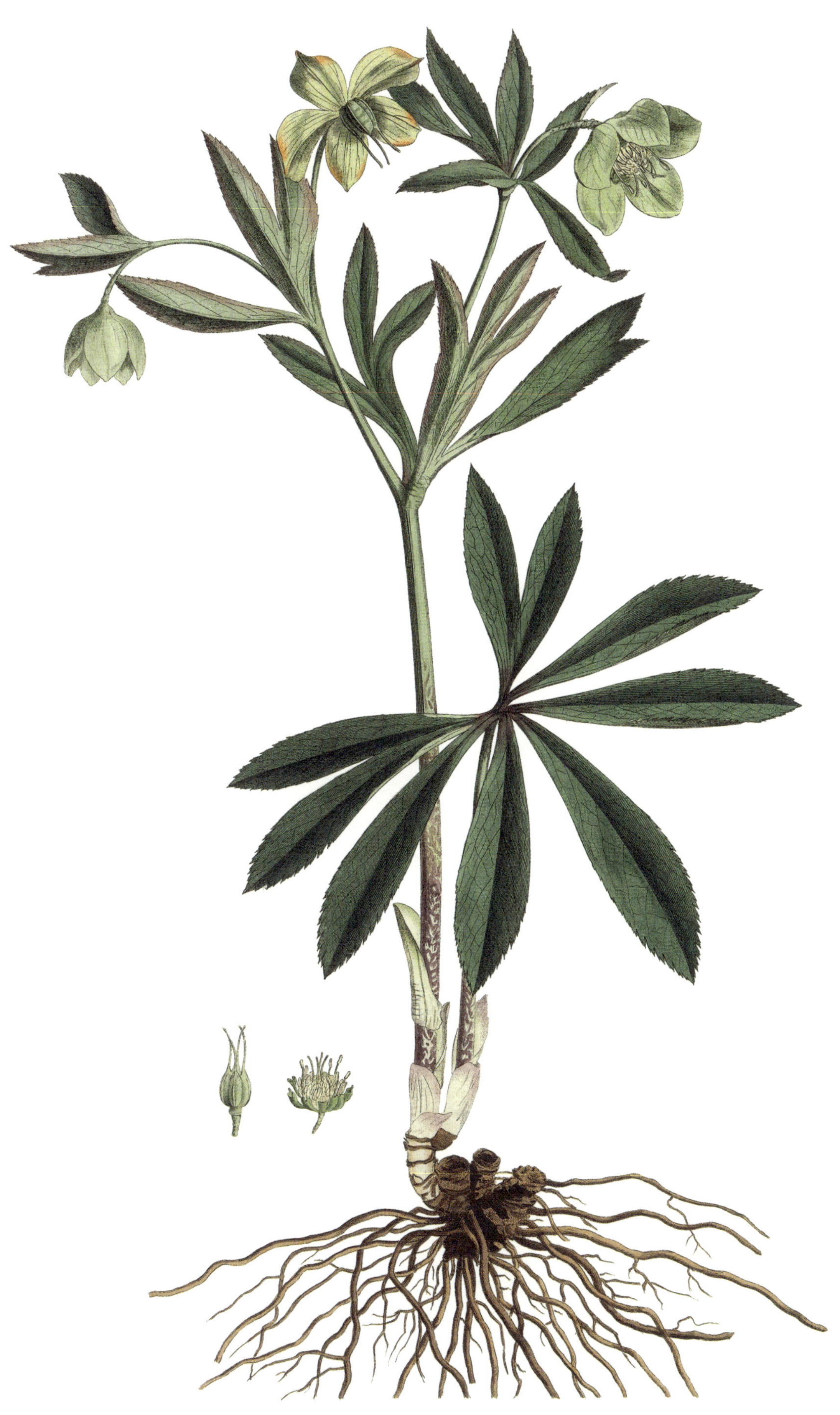

Helleborus viridis

BLACK NIGHTSHADE

Solanum nigrum

For a plant with such a macabre name, in the flesh, black nightshade doesn't seem to fit the bill. Its healthy green foliage, dainty little white flowers and pert berries broadcast innocent good looks. If only the plant wasn't coursing with solanine, a toxin that can do a lot of damage to mammals, including two-legged ones. The chemical is common in nightshades, but it was first identified in the fruits of this very plant. It doesn't trouble birds, which get the best of the berries and unwittingly spread its seeds. But don't have nightmares: it's not a hardy brute. The first cold night kills it off.

ANATOMY NOTES

SEEDS — Black nightshade seeds can survive for around 80 years in undisturbed soil.

LEAVES — The leaves have been used as a poultice for open wounds and infections, but there are much better plants for making herbal plasters.

FRUITS — Concentrations of solanine are greatest in the unripe green berries, but toxicity reduces as they ripen. Botanists argue about exactly how toxic ripe berries are; it's best to give them a miss until they decide. There are some closely related plants, confusingly sometimes also called black nightshade and part of what's known as the *Solanum nigrum* complex, with sweet edible berries used for jams, pies and sauces.

ROOTS — The ability of black nightshade roots to suck up heavy metals from the soil is being investigated for possible land decontamination and reclamation uses.

ESSENTIALS

Aka — Garden nightshade, European black nightshade

Family — Ranunculaceae (buttercup)

Urban habitat — Gardens, cultivated and disturbed land

Height — 90 cm / 3 ft

Flowers — July to September

Uses — Soil decontaminant

Grow — Sow seed in spring

BOTANIST'S WARNING

The toxin solanine stops cells in your body from working properly. It breaks down their membranes and interferes with their mitochondria, the tiny structures inside that generate the energy each cell needs to function and survive. Symptoms of poisoning range from diarrhoea and vomiting to paralysis, fever and death.

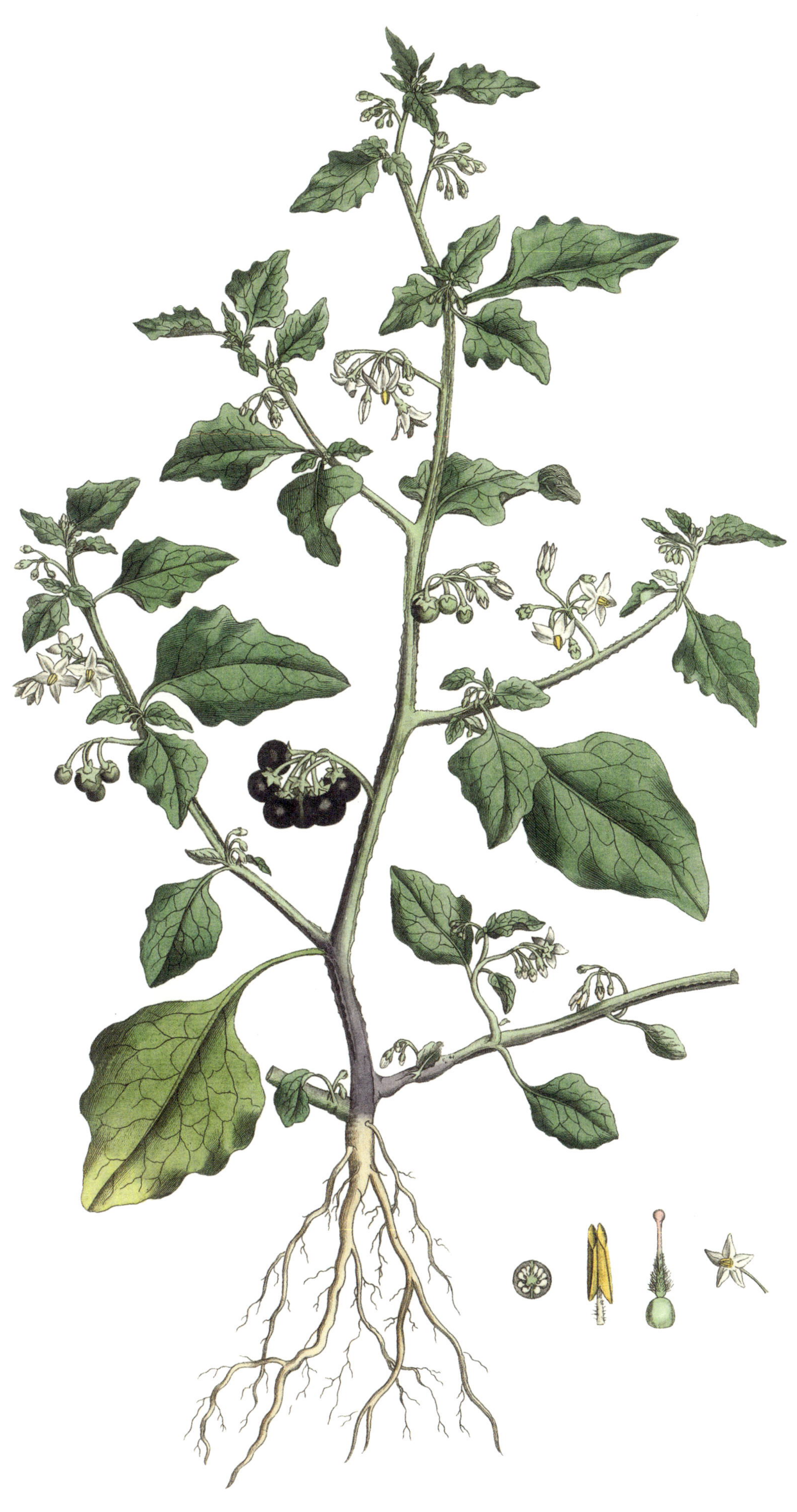

Solanum nigrum

LORDS-AND-LADIES

Arum maculatum

Please be upstanding for lords-and-ladies! Maybe not quite as upstanding as the plant itself, which resembles erect male genitalia – and, bizarrely, female, too. Although all parts of this woodland plant are extremely poisonous, washerwomen braved blistered hands to use its roots for starch to stiffen the ruffs of the Elizabethans and clean communion cloth. Appearing in April, it often goes unnoticed until autumn, when it produces spikes loaded with deadly-looking scarlet berries. They taste revolting, which is a blessing, because eating them can cause pain, breathing difficulties and even death.

ANATOMY NOTES

SPADIX — This phallic protrusion at the plant's centre gives off a faecal odour, unpleasant to us, but highly attractive for the insects that pollinate it.

FLOWERS — Rows of tiny flowers ring the base of the spadix. As insects feast from the female flowers, male flowers release pollen onto them. When they feed at neighbouring plants, the insects then pollinate other female flowers.

ROOT — Although poisonous when fresh, the root can be dried and processed into edible flour or starch, known as sago. This was done on an industrial scale in Portland, Dorset, until about 1850.

LEAVES — The cloak-like leaf encircling the spadix is called a spathe. It protects the reproductive parts of the plant.

BOTANIST'S WARNING

There are very few cases of human deaths by poisoning from lords-and-ladies, but a fair number of cattle have met an untimely end by freely munching the attractive red berries. The plant contains sharp, microscopic crystals (calcium oxalate) that can irritate skin, tissue and organs, and cause potentially lethal swelling in the windpipe.

ESSENTIALS

Aka — Cuckoo pint

Family — Araceae (arum)

Urban habitat — Under hedges

Height — 30 cm / 1 ft

Fruits — Autumn

Uses — Poison, flour, starching fabric

Grow — Scatter the seeds in the autumn

Harvest — Irritant, handle with gloves

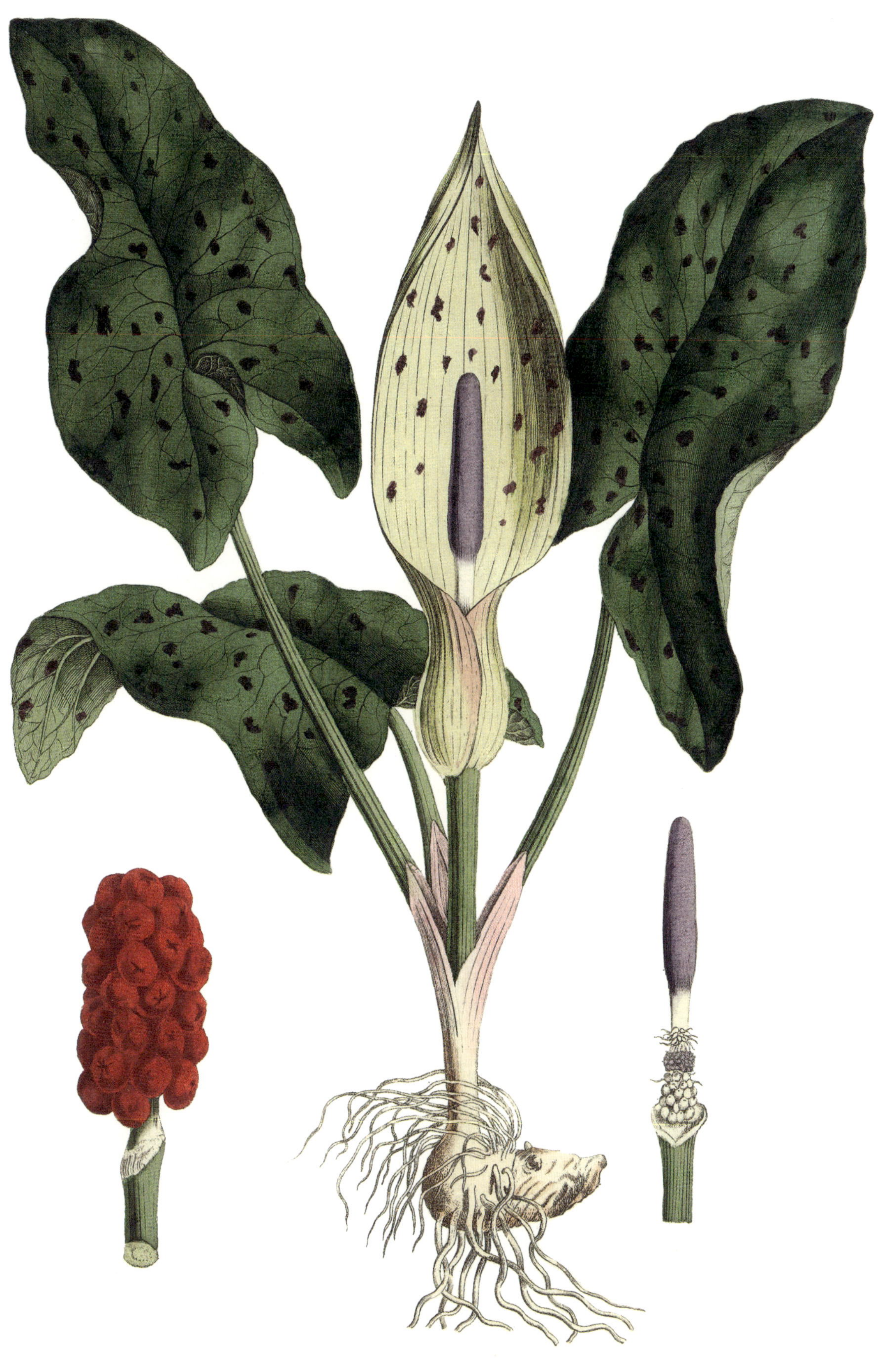

Arum maculatum

WOOD ANEMONE

Anemonoides nemorosa

Wood anemones know just when to act. Preferring undisturbed spots in old woodlands, they burst into flower in early spring before the trees and hedges above them come into leaf and steal their sunlight. Their distinctive star-shaped flowers create dazzling constellations across forest floors, but are now found in parks, gardens and cemeteries about town too. Don't be tempted to pick a bunch; if damaged the wood anemone releases a highly toxic chemical that burns the skin or does much worse if eaten. To the Chinese it's the 'flower of death', but really, it's a beautiful plant with an ingenious means of self-defence.

ANATOMY NOTES

RHIZOMES — The delicate rhizomes (creeping horizontal roots) of the wood anemone split off very easily, making lots of new plants, but it's a very slow process. Large swathes in a forest can indicate ancient woodland.

FLOWERS — A portent of sickness and ill-fortune in some cultures, the Romans nevertheless viewed it as a lucky plant. Picking one early in the season was believed to ward off fever for the year, whereas in old English traditions, the first flower was wrapped in silk to keep pestilence at bay. The flowers provide pots of nectar for early emerging bees; they close up when the weather is inclement to avoid damage to their long stamens.

LEAVES — In old texts, this plant is referred to as wood crowfoot because of the shape of the leaves.

BOTANIST'S WARNING

The wood anemone is a plant that can irritate, burn and even kill. Everything is fine until it's damaged, when a compound called ranunculin contained within its cells turns into protoanemonin. This highly toxic chemical not only produces a smell that scorches the insides of your nostrils, but can literally burn and blister skin when touched. If eaten (admittedly, more likely among grazing animals than humans), it irritates the digestive tract, causing vomiting, diarrhoea, paralysis and, in extreme cases, death. So, wear gloves when handling this plant in the garden.

ESSENTIALS

Aka — Wood crowfoot

Family — Ranunculaceae (buttercup)

Urban habitat — Shady, deciduous woods and hedges

Height — 20 cm / 8 in

Flowers — March to May

Uses — Ornamental plant for shady spots

Grow — Sow seed in autumn

Harvest — Always wear gloves when handling this plant

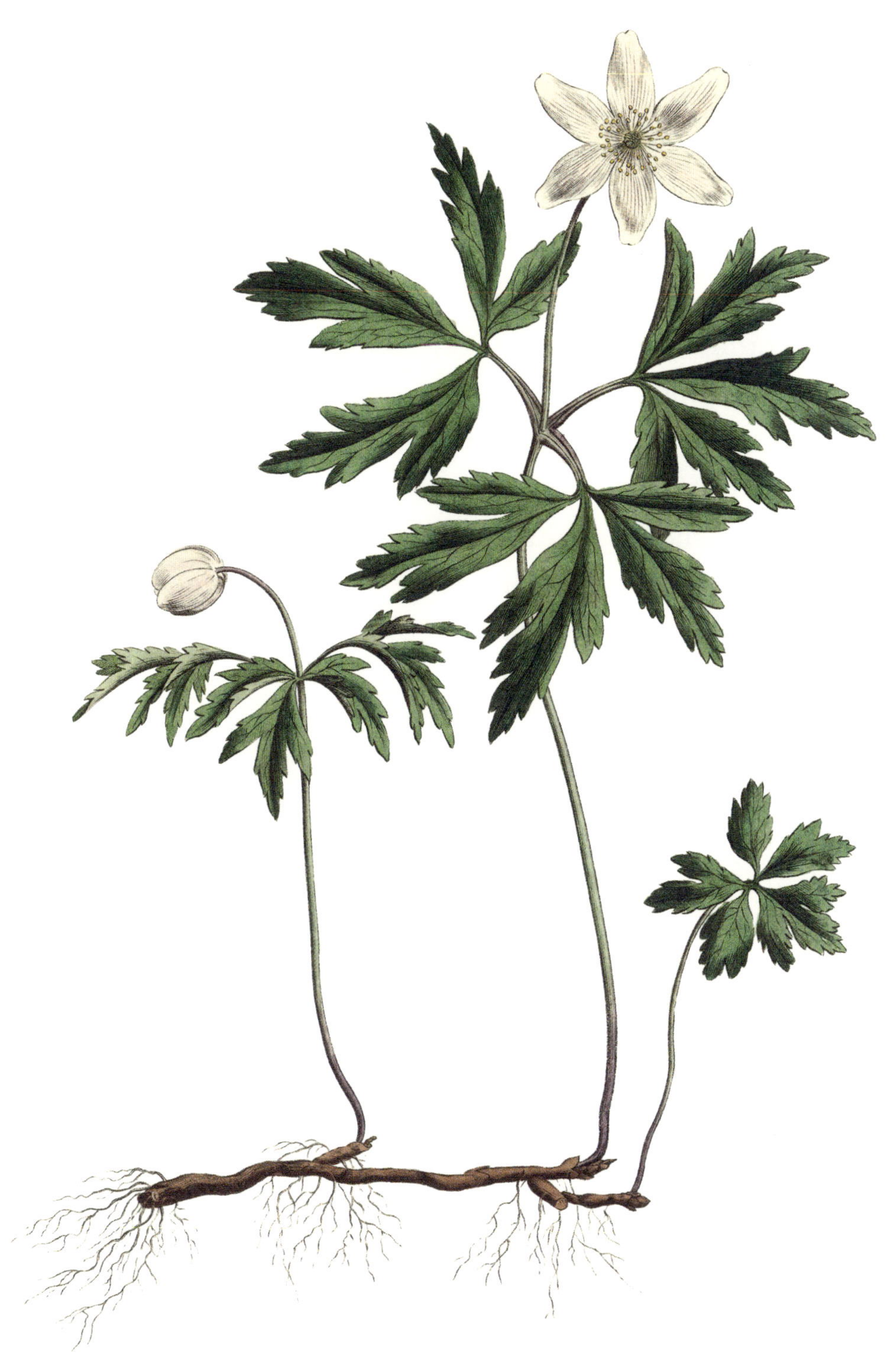

Anemonoides nemorosa

FOOL'S PARSLEY

Aethusa cynapium

The summery white flowerheads of fool's parsley can line pavements and footpaths around town like a miniature weather front of little fluffy clouds. They present quite a feast for insects, especially flies – but definitely not humans. When young, it particularly resembles ordinary garden parsley, its culinary counterpart. But, as the name warns, you'd be a fool to eat fool's parsley. At least you'd know pretty quickly that you'd made a mistake. It burns the insides of your mouth and throat, and can even blister. Keep going and muscle paralysis, vomiting, sensory loss and death await. So leave the eating to the flies.

ANATOMY NOTES

FLOWERS — A posy of fool's parsley may look attractive, but as you might expect of gifting a poisonous plant to a loved one, it traditionally symbolises silliness and foolishness.

LEAVES — Fool's parsley can grow wild in gardens, though it would be bad luck to find it in your parsley patch. Crush the leaves and you can tell them apart; fool's parsley either has no smell or a pretty horrendous one, depending on the point in the lifecycle. Its leaves are also much darker than its edible cousin. The plant has been used in homeopathy to help soothe babies with severe colic who can't keep their milk down. Still, it's a brave mother who administers poison to a newborn.

BOTANIST'S WARNING

Fool's parsley contains toxins called cynopine, which creates the burning sensation in the mouth, and coniine which causes muscle paralysis spreading from the feet upwards. With a strong enough dose, it paralyses the respiratory system, which could kill without urgent intervention. Do stick to garden parsley for your omelette garnish.

ESSENTIALS

Aka — Dog poison

Family — Apiaceae (carrot)

Urban habitat — Gardens, hedges, roadsides, pavement margins

Height — 1.2 m / 4 ft

Flowers — June to September

Uses — Poison, good for pollinators

Grow — Sow seed in autumn

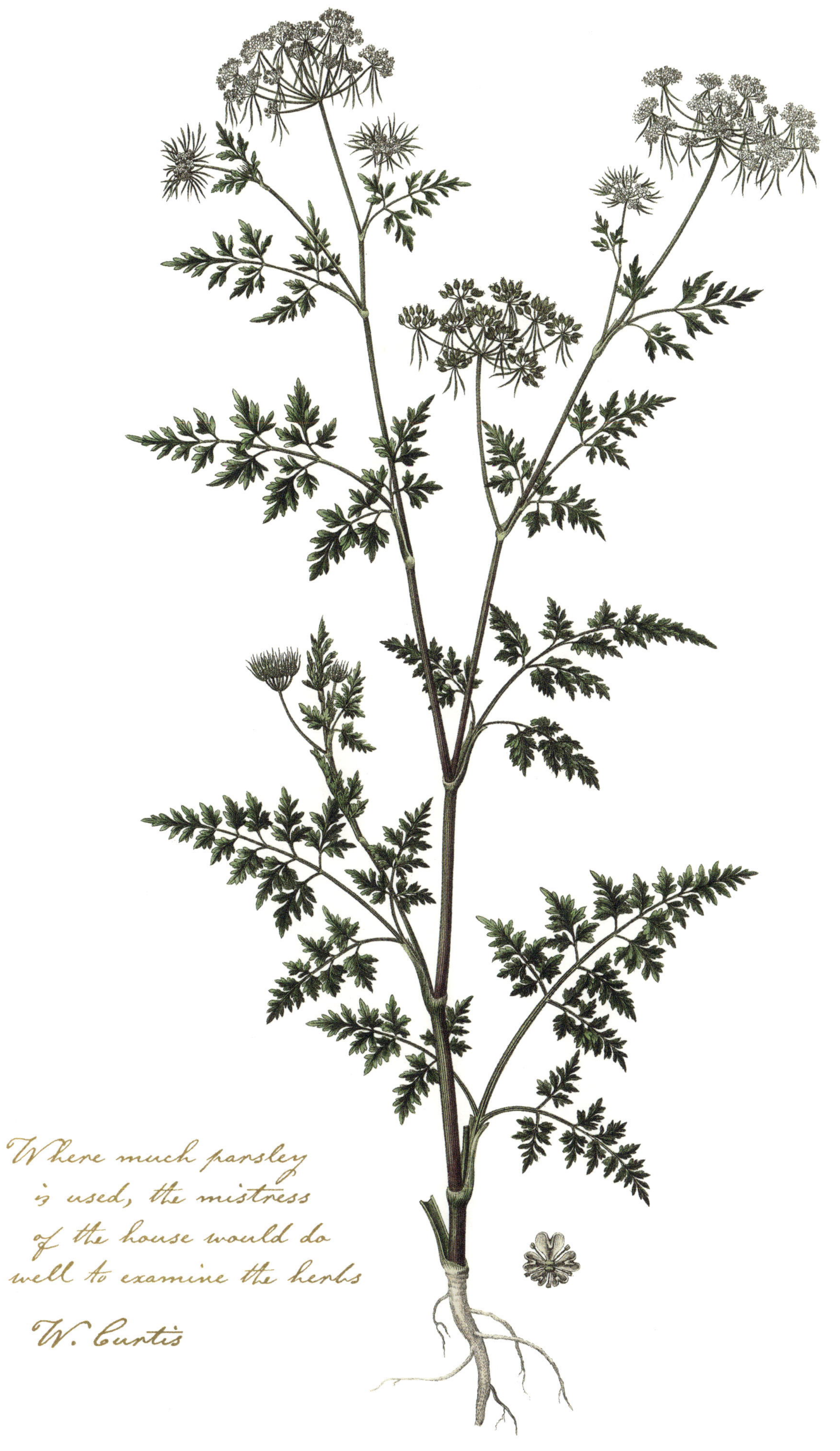

Aethusa cynapium

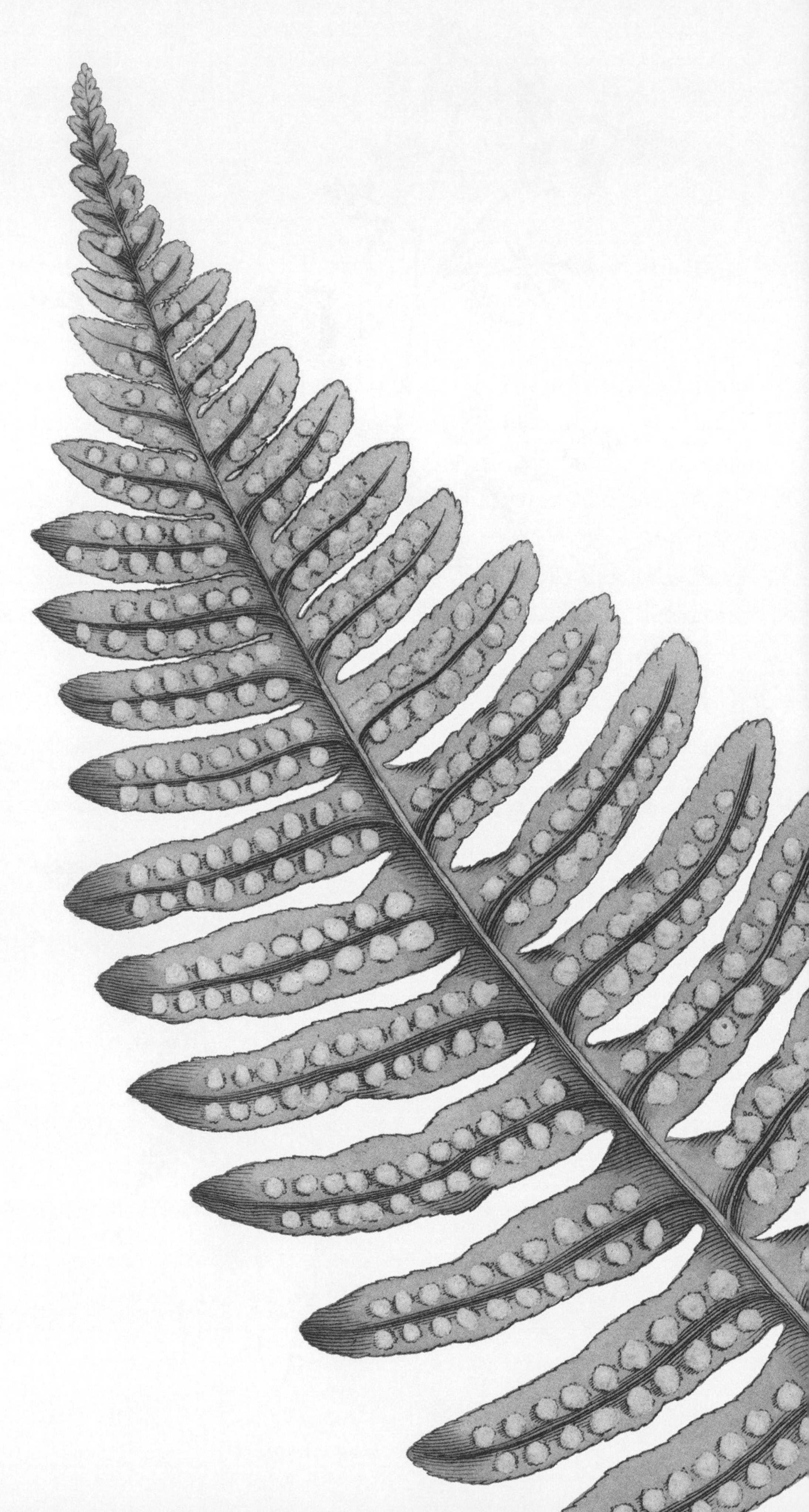

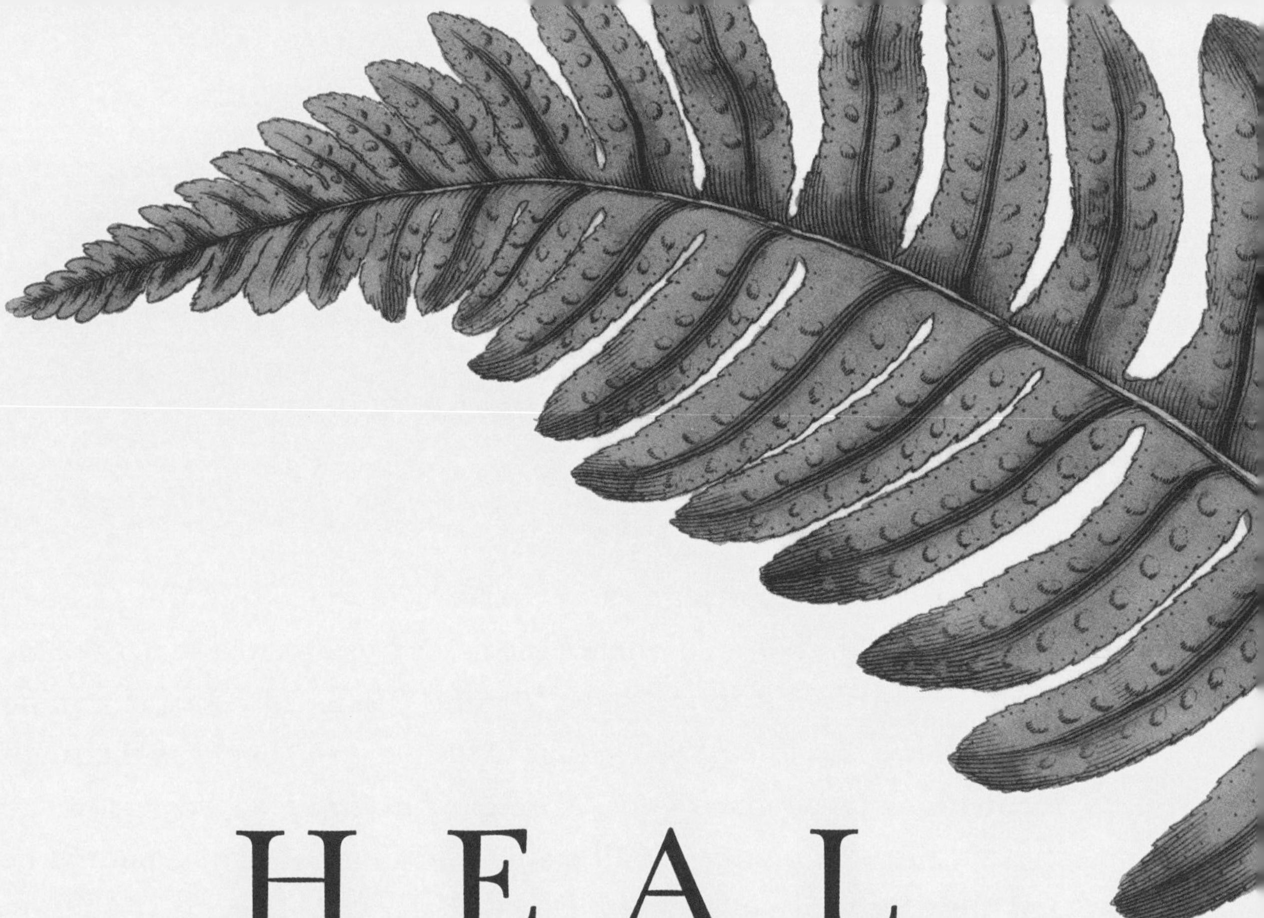

HEAL

Discover *NATURAL REMEDIES*
and find relief from all manner of aches,
agues and *AILMENTS*
in the al fresco *PHARMACY*.

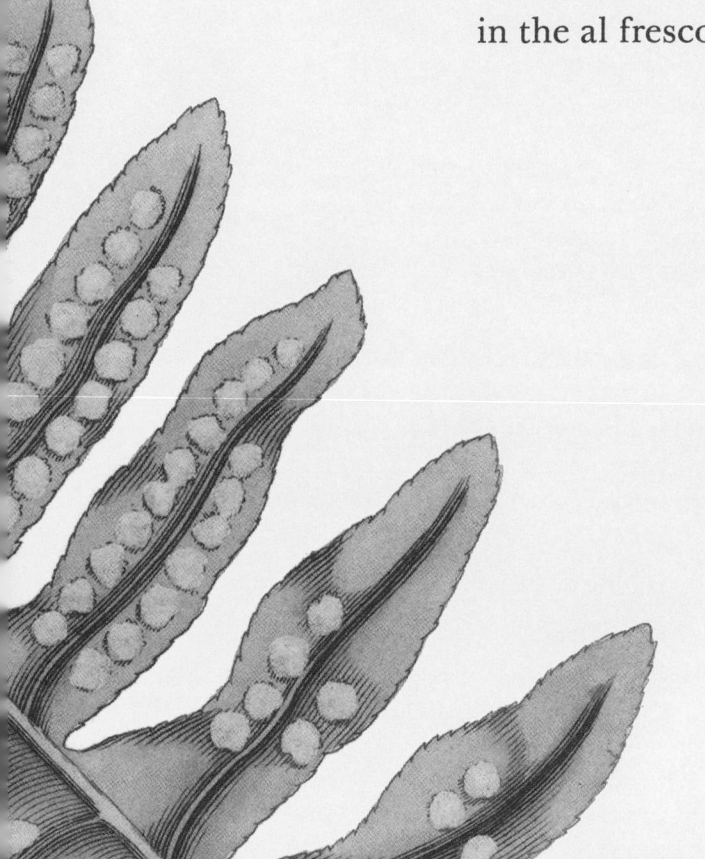

C O W S L I P

Primula veris

The classic flower of pastures and hay meadows – once thought to grow wherever the cows had 'slopped' (hence the name) – suffered a sharp decline in the countryside, thanks to intensive farming. Thankfully, this charming butter-yellow flower has found some refuge in the grass verges of city thoroughfares, even without its bovine friends. Appearing in spring, its flowers can be made into a tea that's a panacea to all manner of winter ills, rejuvenating the mind and body for the year ahead.

ANATOMY NOTES

FLOWERS — Charles Darwin was the first to note that cowslips have two different types of flower, with subtle differences in their reproductive parts. Each plant is all of one type (or morph). The feature is called heterostyly and prevents the plant from fertilising flowers of the same type, discouraging inbreeding. The flowers have several uses, such as adding its gentle citrus flavour to homemade wine and fizz, or as the basis of an anti-wrinkle face wash. The flowerheads look similar to a bunch of keys; one story goes that when St Peter found out the Key of Heaven had been duplicated, he was so shocked he dropped his set. Where these keys landed, cowslips grew, which is how we get the common name, St Peter's wort.

LEAVES — The young leaves are good to eat raw in salads, especially with a dressing to counter their tartness.

ROOTS — The roots contain mild saponins, which were used in herbal medicine against coughs. The root can be dried to be used year-round.

BOTANIST'S REMEDY

TEA — This tea can be used to ease many ailments, including coughs and colds, headaches, insomnia and anxiety. Brew 2 teaspoons of dried or 4 teaspoons of fresh flowers (homegrown is best) in boiling water for 5 to 10 minutes. Strain and enjoy.

ESSENTIALS

Aka — St Peter's wort

Family — Primulaceae (primrose)

Urban habitat — Grassy areas

Height — 30 cm / 1 ft

Flowers — April to June

Uses — Therapeutic tea, edible leaves and flowers, face wash

Grow — Sow seed in autumn

Harvest — Leaves in early spring, flowers in mid spring

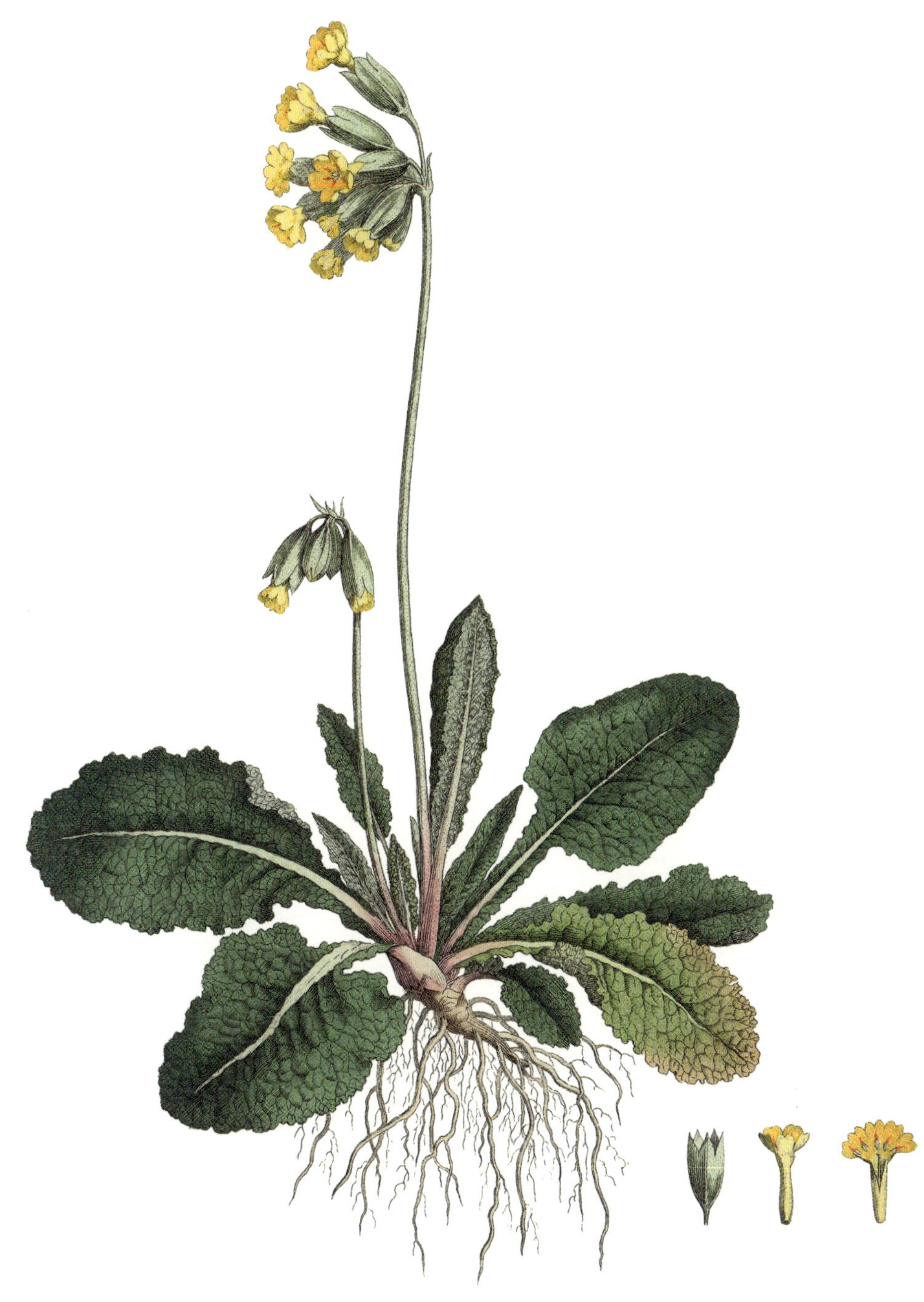

Primula veris

FIELD SCABIOUS

Knautia arvensis

Field scabious's fluffy lilac flowers are supposed to be instantly soothing to the senses – just as well on a hot summer's day stuck in a traffic jam, when you might spot them bobbing about on the roadside. Early herbalists believed their rough, hairy stalks should be used to treat rough, itchy skin, on no better basis than that they looked similar – a principle known as the doctrine of signatures. This time it seems they were right. Once used for treating skin lesions of bubonic plague and scabies, it's still found in today's herbal remedies and commercial products for many kinds of skin complaints.

ANATOMY NOTES

LEAVES — The plant once filled hay meadows and pasture, which was no bad thing as grazing animals loved to chow down on its abundant foliage. Its name comes from the Latin *scabere* 'to scratch'; but field scabious wasn't only to treat itchy skin. Leaf infusions were used for coughs, colds and heart problems, while wine infused with the leaves was intended to relieve pleurisy. Bruised leaves applied to boils for several hours may help heal the infection.

FLOWERS — The gently curving surface of the flowerhead makes a perfect landing pad for pollinators, which you'll see on them in great numbers. In times past, maidens would name several blooms of field scabious after potential suitors and the one which flowered the best would be her husband. Seems about right.

BOTANIST'S REMEDY

SKIN SALVE — Put 130 g / 5 oz of dried flowers (ideally homegrown) and 450 ml / 15 fl oz of base oil (olive oil works well) in a bowl and heat over a pan of simmering water (a bain-marie) for an hour. Strain and keep in a sterilised jar. This infused oil keeps for 6 months. To make the salve, which lasts for a year, add 170 ml / 6 fl oz of the infused oil to 50 ml / 2 fl oz coconut oil and 30 g / 1 oz beeswax in another bain-marie. A microwave also works. Pour into sterilised tins and allow to cool.

ESSENTIALS

Aka — Bachelor's buttons, gypsy rose

Family — Caprifoliaceae (honeysuckle)

Urban habitat — Dry grassy areas and verges

Height — 90 cm / 3 ft

Flowers — July to September

Uses — Skincare

Grow — Sow seed in spring

Harvest — Flowers in summer

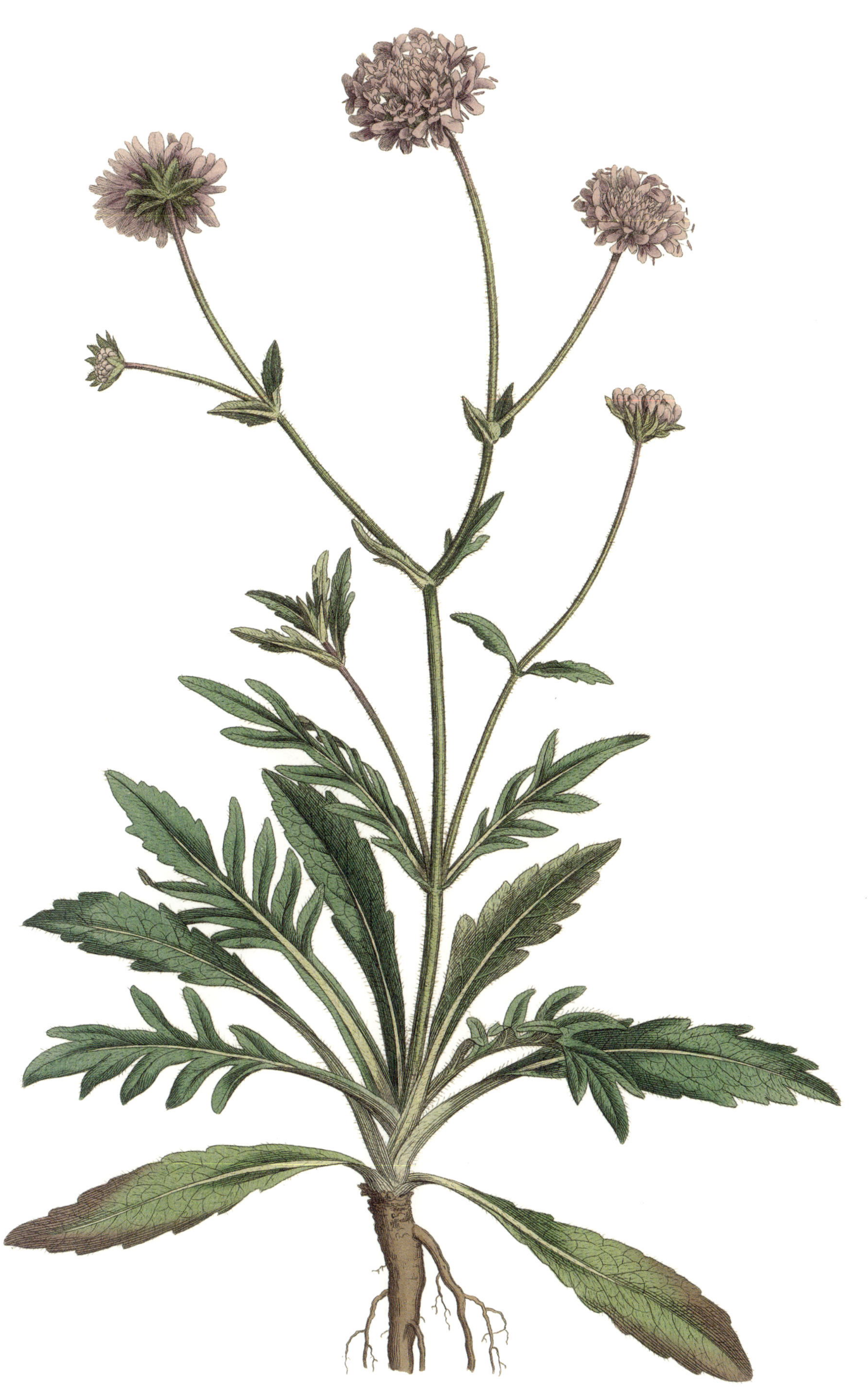

Knautia arvensis

BROAD-LEAVED DOCK

Rumex obtusifolius

If there's one plant that everyone knows as a healer, it's dock – so much so that its name is often confused with 'doc'. Often growing in abundance near its nemesis, the nettle, dock is the go-to sting remedy that makes herbalists of us all. But it's often forgotten that this commonplace weed is also a respectable cooking leaf, best sautéed in butter or gently simmered. Looks-wise, the humble dock isn't going to win any awards, but it should definitely get recognition for the many insects it shelters and feeds, including the small copper butterfly, which is in decline. Here's a weed to celebrate!

ANATOMY NOTES

SEEDS — Dock gets everywhere because of its prolific seed production, a reason it's burdened with the label 'weed'. As many foragers say, the best way to combat weeds in your garden is to eat them. The seeds are nutritious and can be ground to add to flour for baking.

LEAVES — The leaves are best eaten young, but they're quite bitter raw, so cook them. They're good with butter; in fact, the leaves were once used to wrap butter, because they can be so capacious, hence its local name, butter dock. In Romania, dock is sometimes used instead of cabbage in the traditional dish of sarmale, cabbage rolls.

ROOTS — A tea made from the roots is a traditional herbal remedy for respiratory problems, constipation and jaundice.

BOTANIST'S REMEDY

STING-RELIEF TREATMENT — Anyone who enjoys a ramble can at some point expect the inevitable – a nettle sting. Dock is the free and instant remedy. For best results, crush the leaf, making sure to open its large mid-rib and smaller veins to get plenty of sap on the affected area. Some claim the sap contains soothing natural antihistamines, but the constituent parts of dock still aren't well understood. Some think the evaporation of the sap cools the sting, while others say it's just a placebo effect. Whoever's right, all these theories agree that dock does indeed help.

ESSENTIALS

Aka — Dock leaf, butter dock

Family — Polygonaceae (knotweed)

Urban habitat — Widespread in grassy areas, woods and gardens

Height — 1 m / 3 ft 4 in

Flowers — July to October

Uses — Soothes stings, edible leaves and seeds

Grow — Plant seeds in spring

Harvest — Young leaves in spring for cooking, seeds in autumn

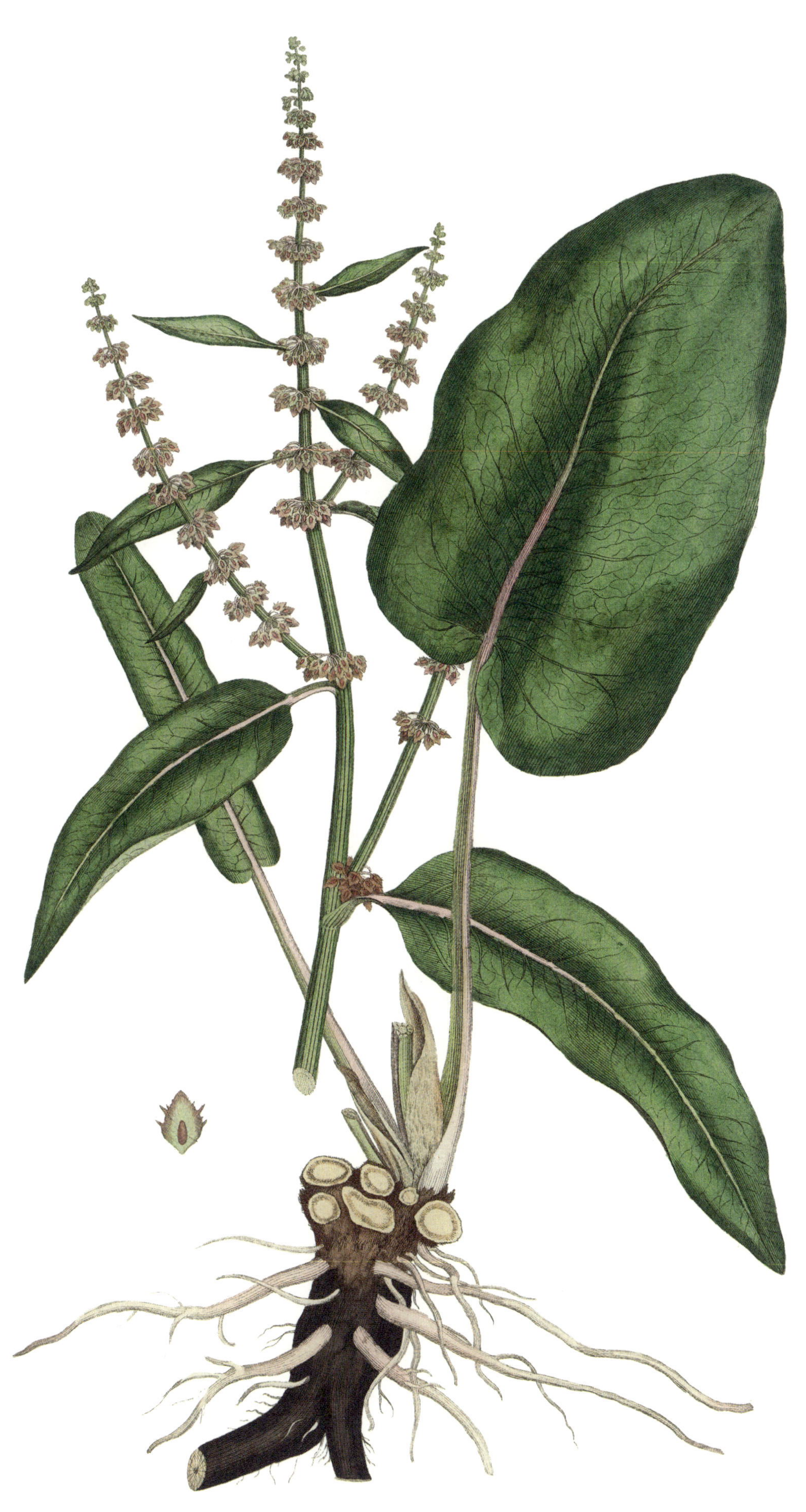

Rumex obtusifolius

FUMITORY

Fumaria officinalis

Clouds of this scrambling plant hang at the edge of the road like low-lying smoke, an illusion made by its misty grey-green leaves and almost luminous pink and crimson-tipped flowers. It's been associated with smoke for a long time; even the Roman naturalist Pliny the Elder mentioned it. Fumitory is an apothecary plant (indicated by the *officinalis* in the scientific name), and has been used by herbalists since antiquity to treat a lengthy list of complaints, from arthritis to impure blood. These days, it's still used as an active ingredient in some commercial skincare products.

ANATOMY NOTES

FLOWERS — Herbalists used dried flowers to make a digestive tea that could stop the hiccups; the cooled tea was also used as cleansing face wash. The purple flowers produce a warm yellow dye for textiles and cosmetics.

LEAVES — Historical medicinal uses were wide-ranging, from treating liver and spleen problems to making a gargle for sore throats. The leaves were once burnt to ward off evil, but a more up-to-date use of fumitory is as a rennet to curdle milk to make cheese. Do note, however, that fumitory is poisonous and no longer recommended for home use.

STEM — The stem contains a latex sap which is poisonous when raw.

ROOTS — Pulling the roots from the ground can release a gassy smell. This probably encouraged the ancient belief that the plant didn't grow from seeds, but came from natural smoke held in the earth. Its medieval Latin name, *fumus terrae* (smoke of the earth) is the origin of our name, fumitory.

BOTANIST'S REMEDY

As it can be toxic, fumitory isn't recommended for home remedies. Since the 1950s, however, chemists have used compounds of fumaric acid, which is found in the plant, to treat psoriasis. Unfortunately, stomach upsets are common among the side effects, so more research is needed to make it a better treatment.

ESSENTIALS

Aka — Wax dolls

Family — Papaveraceae (poppy)

Urban habitat — Grassy verges, waste ground, gardens

Height — 10 cm / 4 in

Flowers — May to September

Uses — Cosmetics, tea, face wash

Grow — Sow seed in autumn

Harvest — Leaves in spring

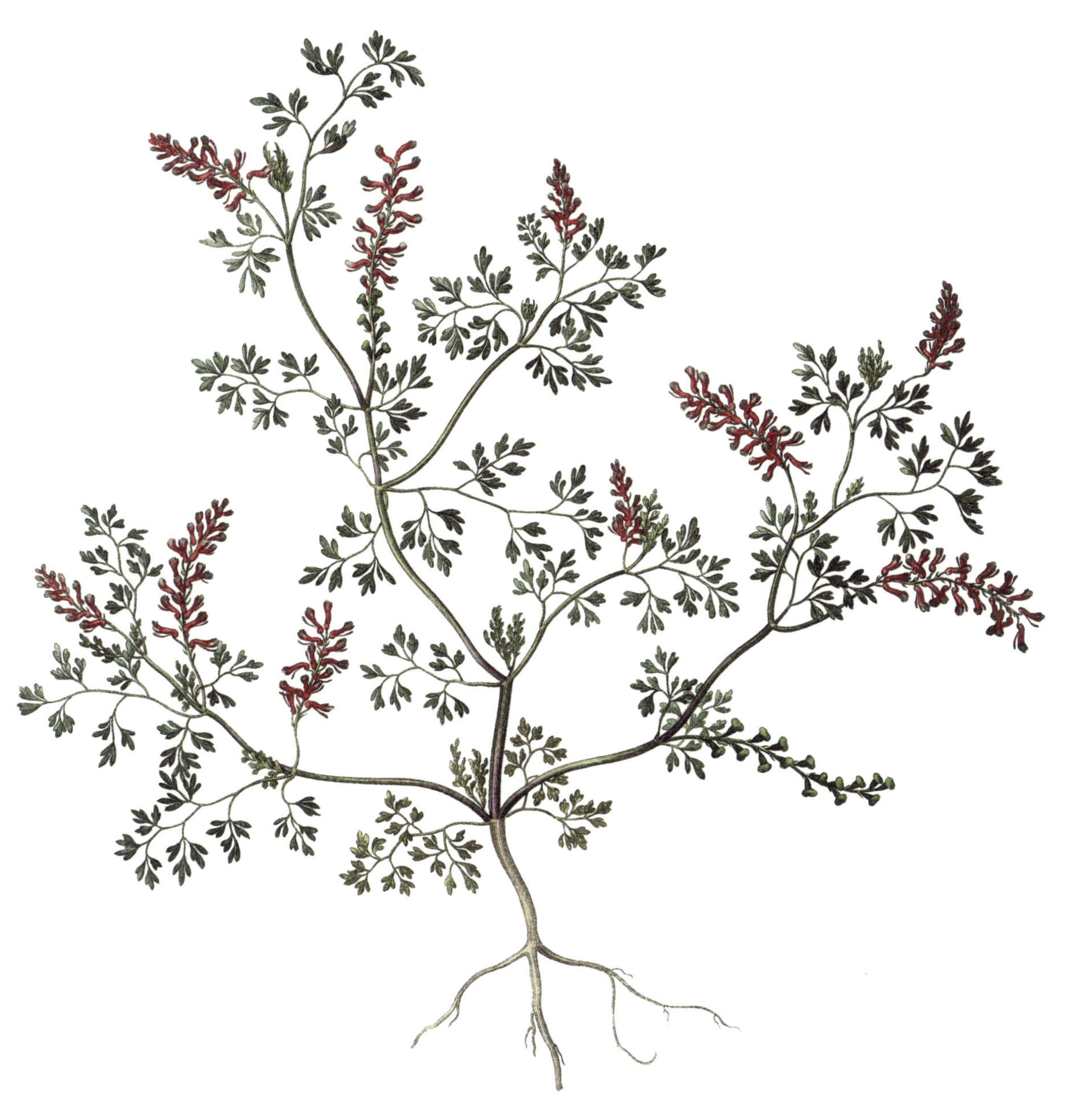

Fumaria officinali

COMMON POLYPODY

Polypodium vulgare

This little evergreen fern streaks walls and pavements with vivid colour, as if Mother Nature has gone berserk with a highlighter pen. Although small, it's pretty handy if you have a sore throat. The roots are surprisingly sweet when made into a tea and exceptionally soothing for a scratchy larynx. It's so good, in fact, that before you realise, you'll be quaffing it by the potful (taking into account its mild laxative properties!). The root is best harvested in autumn; if you dry it out you can enjoy it throughout the year.

ANATOMY NOTES

ROOTS — The roots contain osladin which is at least 500 times sweeter than sugar. It can be used as a natural sweetener, and is sometimes used commercially in nougat. The tea is also said to increase libido in men. In very damp woods, the plant can operate as an epiphyte – surviving on tree branches and getting moisture and nutrients from the air and its surroundings rather than the soil or the tree. In Norway, a gruel made from the plant was popularly eaten before May Day, in the belief it would give protection against snakebite for the rest of the year.

SPORES — The spores form twin rows of striking orange bumps on the underside of the leaves. When mature, they're released on the wind. Patient propagators can collect them in an envelope for sowing in sterile compost. With humidity and heat, they should germinate after several months.

BOTANIST'S REMEDY

TEA — Drink or gargle this tea for a sore throat, using clean fresh root in autumn or dried root at other times of year. Cut it into small pieces and use 2 to 3 teaspoons per cup. Add boiling water and steep for 5 minutes before straining.

ESSENTIALS

Aka — Adder's fern, golden maidenhair

Family — Polypodiaceae (polypody)

Urban habitat — Damp, shady cracks

Height — 30 cm / 1 ft

Flowers — no flowers, but evergreen so can be spotted year-round

Uses — Tea, edible sweet root

Grow — Buy plants to place out in spring

Harvest — Root in autumn

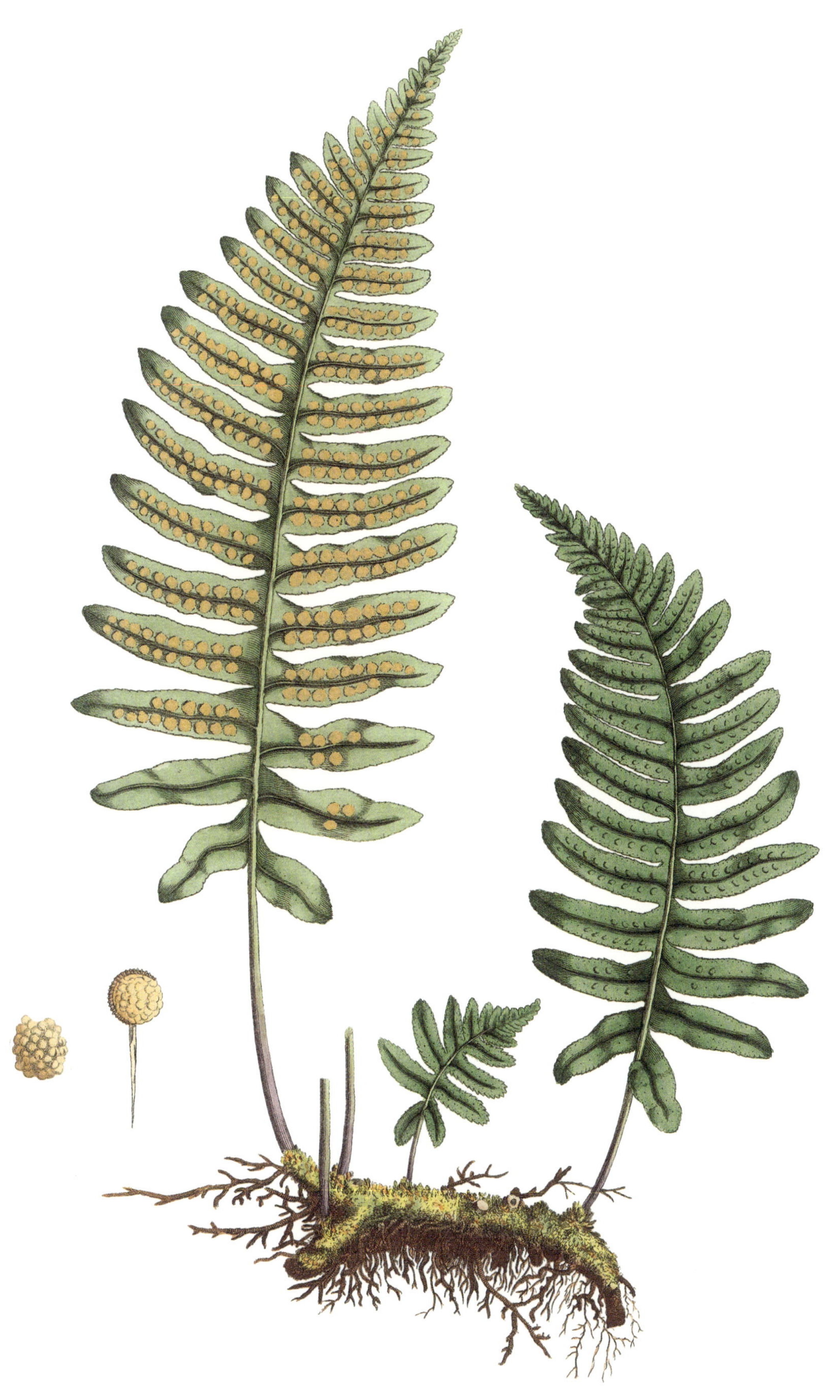

Polypodium vulgare

DAISY

Bellis perennis

She loves me, she loves me not. Daisy chains. The common daisy, its friendly little flowers carpeting parks and gardens just as the weather improves in spring, is often the first plant we learn the name for, the first we pick, the first we play with. As if completely designed for children (in Scotland it's known as bairnwort), it's superb to heal fresh wounds and bruises, and can be picked and applied in the playground, there and then. An essential part of the urban first-aid kit.

ANATOMY NOTES

FLOWERS — As you pluck daisy petals, reciting, 'He loves me, he loves me not', you're actually removing an individual flower every time. The flowerhead, small as it is, comprises lots of little flowers, or 'florets', both in the central yellow eye and the surrounding white petals. The white florets are female only, whereas the internal yellow ones are hermaphrodite and self-pollinate.

LEAVES — Infusions of the leaves were used for centuries for fevers, liver disease and many other ailments. Doctors in the Roman army soaked bandages in daisy juice before battle, ready to apply to wounds. You can eat the leaves in salads, along with the flowers. They're a little bitter, so possibly best enjoyed with other leaves, or sautéed in butter with herbs and a dash of balsamic vinegar. Note that some people are allergic to daisies.

ROOTS — The products of boiling the roots (in a decoction) can be used to treat eczema and scurvy. Scientists are investigating the use of daisy roots, leaves and other extracts for HIV therapy.

ESSENTIALS

Aka — Bruisewort, bairnwort

Family — Asteraceae (daisy)

Urban habitat — Common on grass and verges

Height — 10 cm / 4 in

Flowers — April to October

Uses — Heals wounds and bruises, edible

Grow — Sow seeds in spring

Harvest — As required

BOTANIST'S REMEDY

BRUISE TREATMENT — To help heal a bruise, pick a handful of fresh daisy leaves as soon as possible after injury. Crush or tear the leaves and place on the affected area, holding in place with a bandage if possible, or any piece of material you have to hand. Leave on for several hours and repeat if necessary.

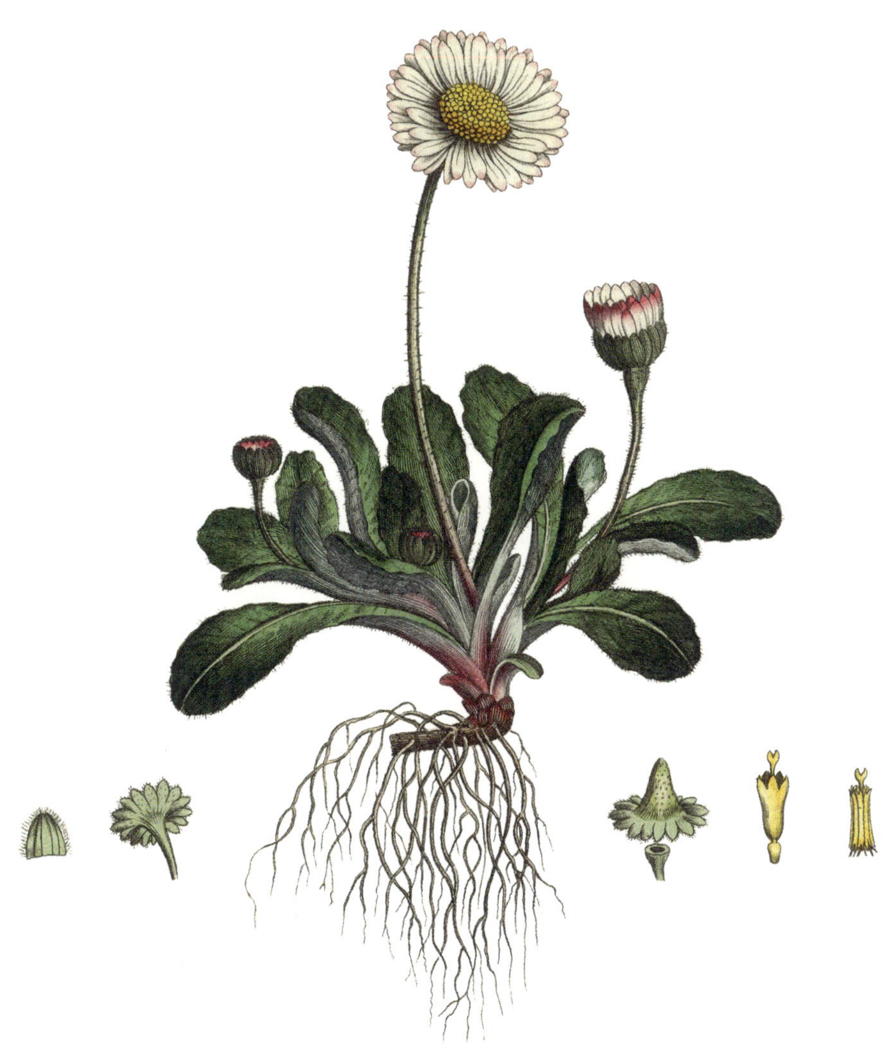

Bellis perennis

GERMAN CHAMOMILE

Matricaria chamomilla

On a hot summer's day in the city, the daisy-like flowers of German chamomile bring instant serenity. It springs up in all kinds of corners – cracked pavements to waste ground – its flowerheads filling the air with a soothing scent, something like a mix of warm earth and honeyed apples. The tea is similarly flavoursome and calming, aids digestion and promotes a good night's sleep. As an antibacterial, antifungal, antiviral and anti-inflammatory too, it's hard to think of a drawback to a mug or two of this delicious floral brew.

ANATOMY NOTES

FLOWERS — If you have blonde hair, chamomile tea is an excellent cold rinse to give it a lustrous sheen – but anyone can use it as a body wash. The flowers are truly multifunctional: they're added to commercial skin cosmetics not just for their fragrance, but as an anti-allergen; they produce a golden yellow dye for textiles; they're good as a natural insect repellent; and they've historically been used in love potions.

LEAVES — The leaves are edible and taste best in early spring, added to a salad, or even to decorate a cake for their subtle floral note. The essential oils extracted from the leaves and flowers are used in many modern perfumes.

BOTANIST'S REMEDY

TEA — Chamomile tea can be made with fresh or dried flowers. When drying flowers, either lay them in the sun for a week, or dry for 8 hours in an oven on its lowest setting with the door open. If they're not fully dry you may lose the whole batch to mould in storage. The tea is simple to make: use 1 or 2 teaspoons of flowerheads per cup, add just-off boiling water and brew for 5 minutes. A lavender sprig in the mix may set you up for an even better night's sleep.

ESSENTIALS

Aka — Scented mayweed

Family — Asteraceae (daisy)

Urban habitat — Warm, sunny spots

Height — 30 cm / 1 ft

Flowers — June to August

Uses — Tea for digestion and calm

Grow — Sow seed in spring

Harvest — Flowerheads in summer

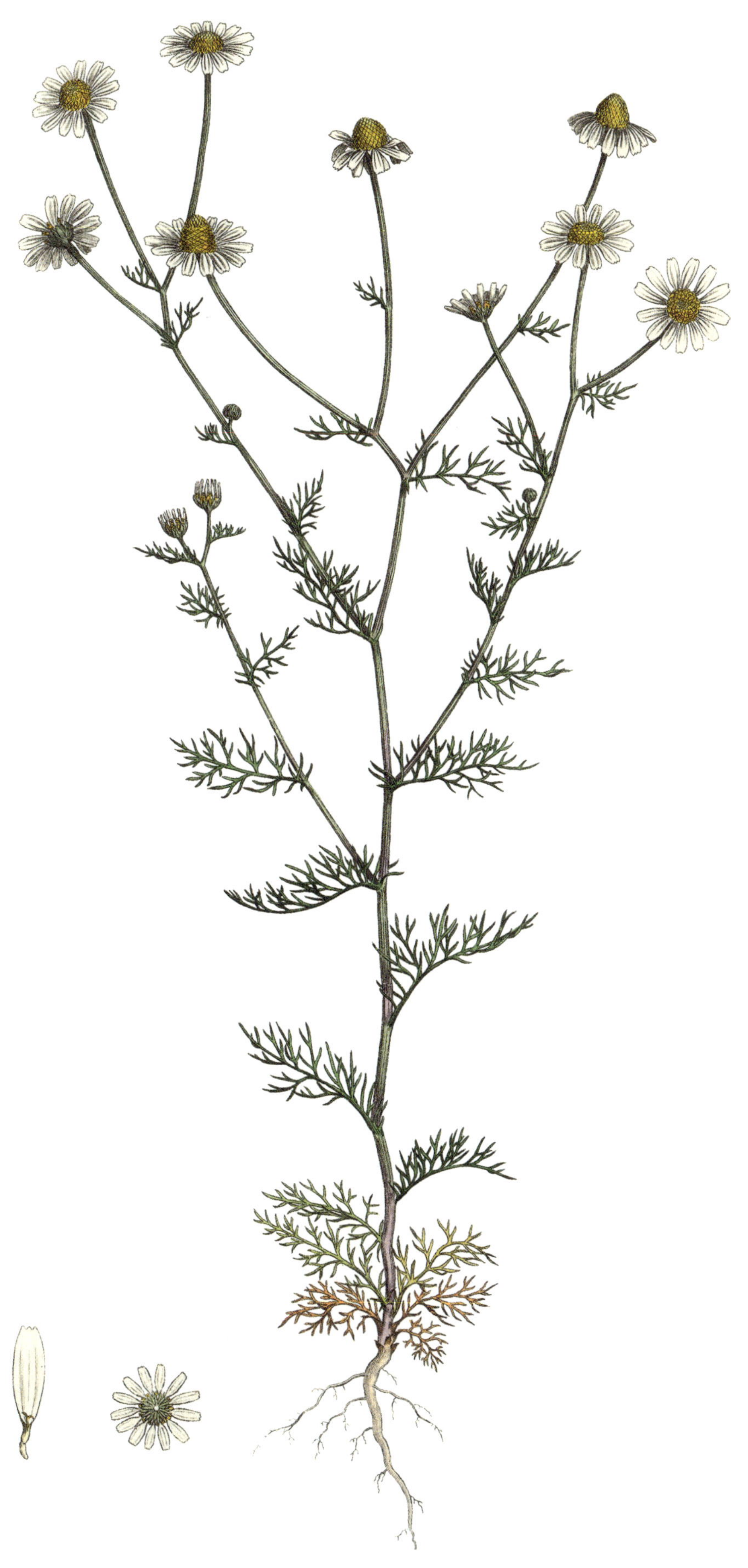

Matricaria chamomilla

MILK THISTLE

Silybum marianum

White veins marble the leaves of the milk thistle like splashes of cream on a hipster's macchiato. No offence to baristas, but after a heavy night out it's actually a nice hot cup of milk-thistle tea you should be reaching for, rather than a strong coffee. Boosting liver function, this is the herbal remedy for over-indulgers. The earthy, antioxidant-rich brew combats liver maladies from cirrhosis to jaundice and kickstarts the digestive system to get the body back in shape. It's great for your insides – just mind your hands on the spines when harvesting.

ANATOMY NOTES

SEEDS — The best and most potent milk-thistle tea is made from dried seeds. If tea's not your thing, you could roast them for adding to caffeine-free coffee substitutes. Some people make oil from the seeds, which has similar health properties.

FLOWERS — Negotiate the merciless spikes on the flowerhead, and you can eat them as a meagre, rather fleshless globe artichoke. As insects do better out of the flowers, it's best to leave them for the pollinators.

LEAVES — The leaves are edible and similarly good for the liver, but do remove the spines first.

ROOTS — Some foragers eat the roots raw or boiled. Young roots can be dried and made into tea too.

BOTANIST'S REMEDY

HANGOVER TEA — Grind dried milk-thistle seeds with a pestle and mortar or blast them in a blender. Gently simmer 2 teaspoons in a mug's worth of boiling water for a few minutes and allow to steep for a further 15 minutes. Strain and drink. A little lemon verbena in the mix brings a citrus lift to help your morning-after rejuvenation.

ESSENTIALS

Aka — Blessed thistle

Family — Asteraceae (daisy)

Urban habitat — Waste ground, verges, cultivated land

Height — 1.5 m / 4 ft 11 in

Flowers — July to September

Uses — Good for the liver

Grow — Sow seed in late spring

Harvest — Roots in spring, seeds in autumn; wear gloves

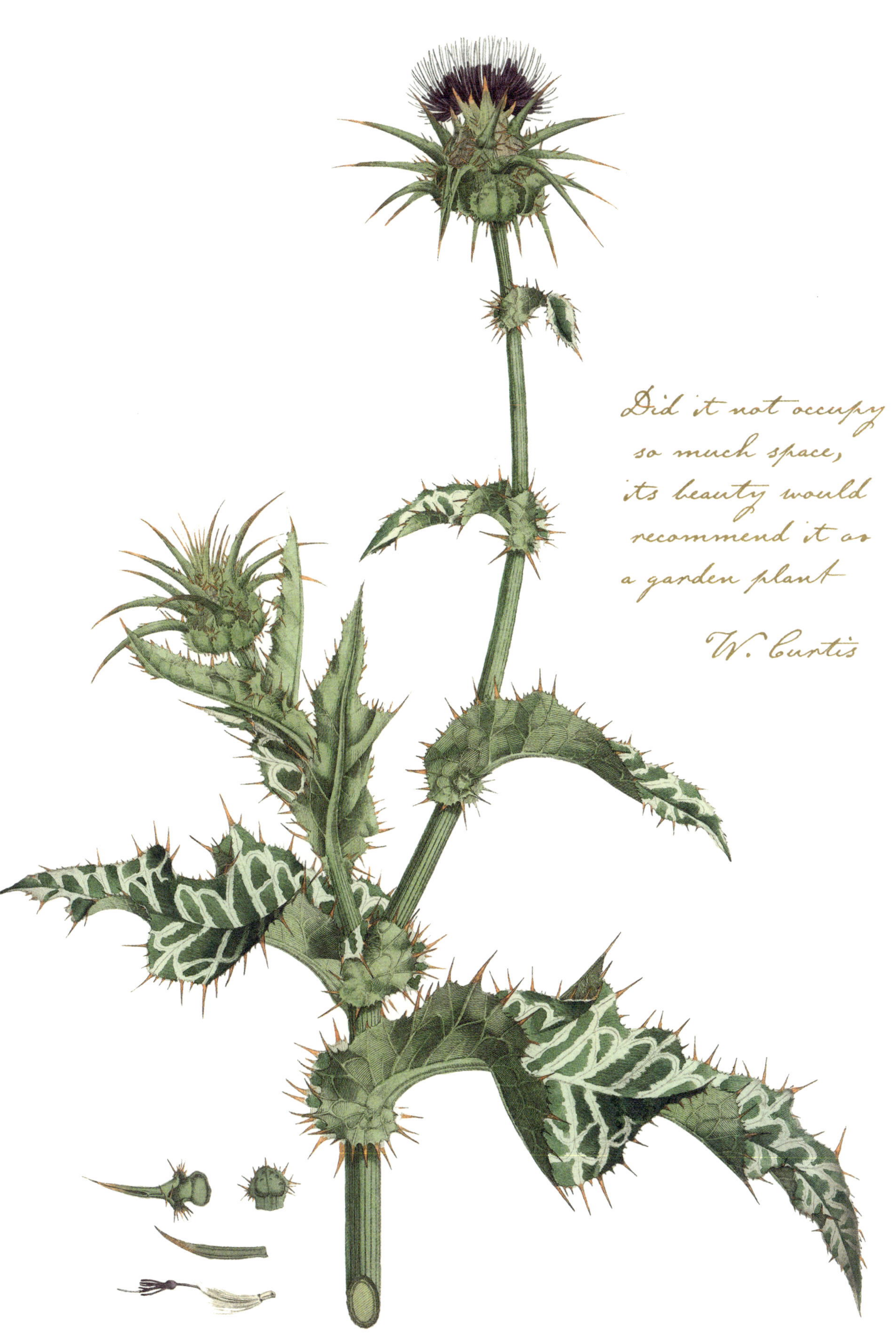

Silybum marianum

WILD VALERIAN

Valeriana officinalis

The soothing properties of valerian have been highly prized since ancient times – although, to this day, no one's sure exactly how the plant's chemical compounds work. The godfathers of medicine, Hippocrates and Galen, raved about the sedative powers of its root extract, which doctors have used ever since to relieve insomnia, epilepsy, muscle cramps and a host of nervous disorders. Medieval herbalists believed it brought peace to 'men who begin to fight', and in both World Wars it was taken for shell shock. Its extraordinary remedial properties are recognised in its Latin name, from the verb *valere* (to be healthy), and *officinalis*, a term only given to the most important medicinal plants.

ANATOMY NOTES

STEMS — Horizontal stems, called stolons, run along the ground forming new shoots and roots, and ultimately perfect clones of the parent plant. They allow wild valerian to spread quickly, making it easy to cultivate.

FLOWERS — The flowers emit a strong sweet scent, a little like vanilla. Essential oil made from the flowers was used in perfumery.

ROOTS — In contrast to its sweet-smelling flowers, the root has a pungent odour reminiscent of cheesy feet. It's dried to make teas, tinctures, oils and extracts for medicinal and therapeutic use. It can also be used as substitute for catnip or bait for rat traps.

SEEDS — Wild valerian is an excellent coloniser thanks to its fluffy 'parachute' seeds that carry easily on the wind.

LEAVES — Rich in phosphorous, the leaves can boost bacterial activity in composts and can be used as the basis of a potent liquid plant feed.

BOTANIST'S REMEDY

TEA — Wild valerian tea, best drunk before bedtime, soothes anxiety and encourages a good night's sleep. Thoroughly wash 2 g / 0.1 oz wild valerian root (dried or fresh) and chop it into small pieces. Pour on 200 ml / 7 oz of boiling water and leave for 15 minutes. Strain out the root pieces and add honey to taste.

ESSENTIALS

Aka — All-heal, common valerian

Family — Caprifoliaceae (honeysuckle)

Urban habitat — Dampish soils near water

Height — 1.5 m / 5 ft

Flowers — June to July

Uses — Medicinal, sedative, analgesic

Grow — Plant out seedlings in spring

Harvest — Roots in autumn of second year

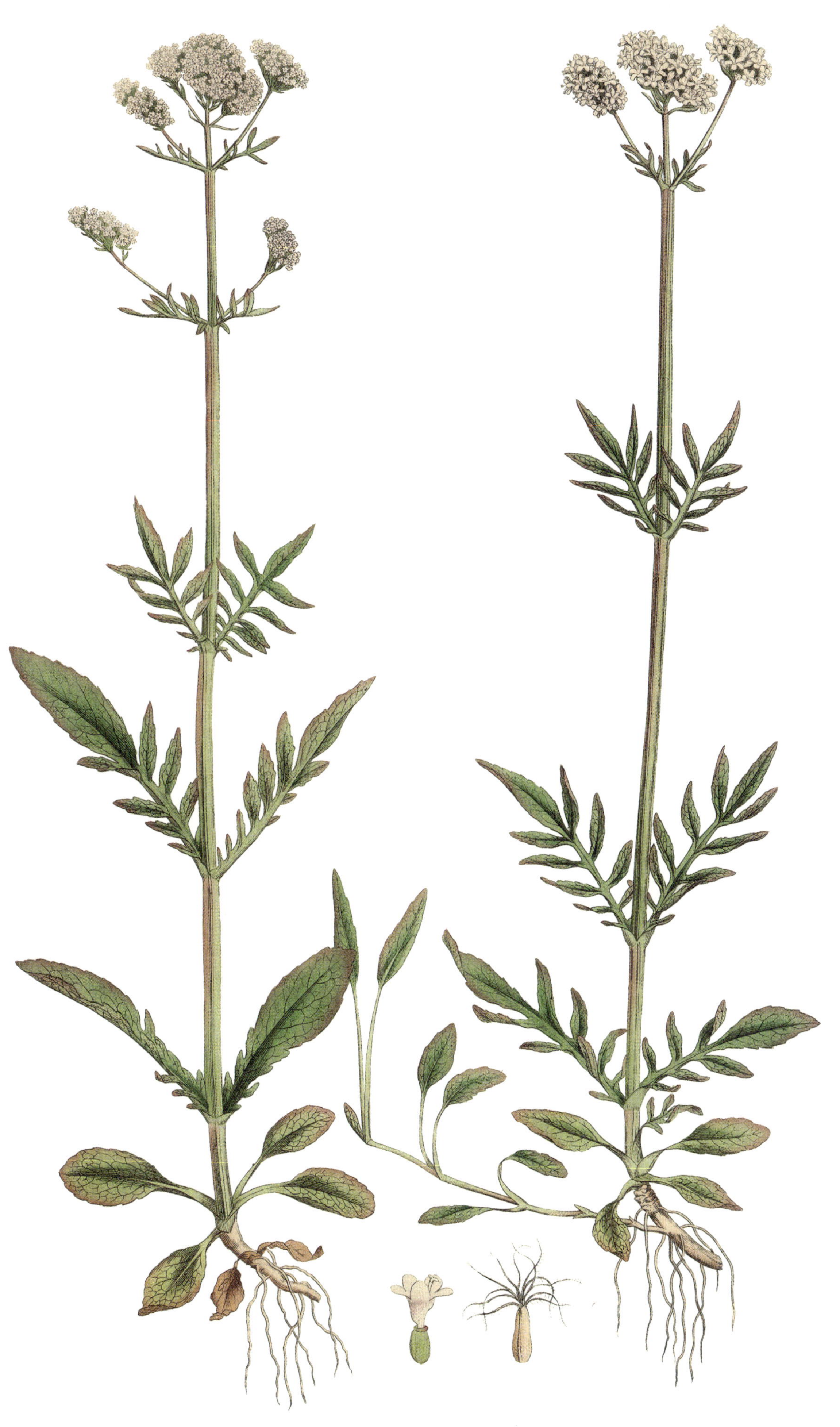

Valeriana officinalis

ROSEBAY WILLOWHERB

Epilobium angustifolium

Favouring scarred, broken and disturbed ground, rosebay willowherb is often the first plant to appear in the wake of a calamity. Its elegant pink cones swiftly colonised London's ruined streets after the Blitz, becoming a symbol of resilience and later the capital's county flower. A true pioneer, it aids the recovery of damaged soils, preparing the way for other plants to follow. But that's just one of its many uses. Russians make the leaves into a restorative tea, Ivan chai, replete with healing and soothing powers. Hitler is said to have feared it so much, he tried to destroy Russian tea factories to undermine the Red Army.

ANATOMY NOTES

SEEDS — Dry out the fluffy seed heads and you have excellent tinder to start your campfire. You can also mix them with feathers to make your duvet warmer.

FLOWERS — The flowers can be made into delicious syrups and jellies; or rubbed on to clothes to help make them waterproof.

LEAVES — Young leaves spruce up a foraged salad and shoots can be lightly steamed or grilled on the barbeque. The leaves have a very unusual feature in that the veins don't extend to the edges but loop around to each other.

STEM — The stem can be made into thread or cord. The insides of the stem, known as the pith, can be dried and powdered and applied to skin to defend against the cold.

ROOTS — As well as being edible once cooked, the fresh roots can be peeled and bruised to use as a poultice on burnt skin.

BOTANIST'S REMEDY

IVAN CHAI — Advocates of this fermented caffeine-free tea swear it's a health wonder. An anti-inflammatory, it's supposed to aid digestion, relieve insomnia and boost energy and immunity. Pick young leaves in spring before the plant flowers. Crush and weigh them down in a sealed jar. Leave for a week, releasing fermentation gases every day, and making sure it still smells good with no mould. Thoroughly dry out after a week. The tea is better with very hot rather than boiling water. Let it brew for at least 5 minutes and strain. Add honey to taste.

ESSENTIALS

Aka — Bombweed, fireweed

Family — Onagraceae (evening primrose)

Urban habitat — Disturbed ground, verges, wood clearings, wasteland

Height — 1.5 m / 4 ft 11 in

Flowers — July to September

Uses — Medicinal, cordage, land management, food, tea, ornamental

Grow — Sow seed in spring

Harvest — Leaves in spring

*The down of the seeds,
mixed with beavers' hair,
has been manufactured
into several articles
of clothing*

W. Curtis

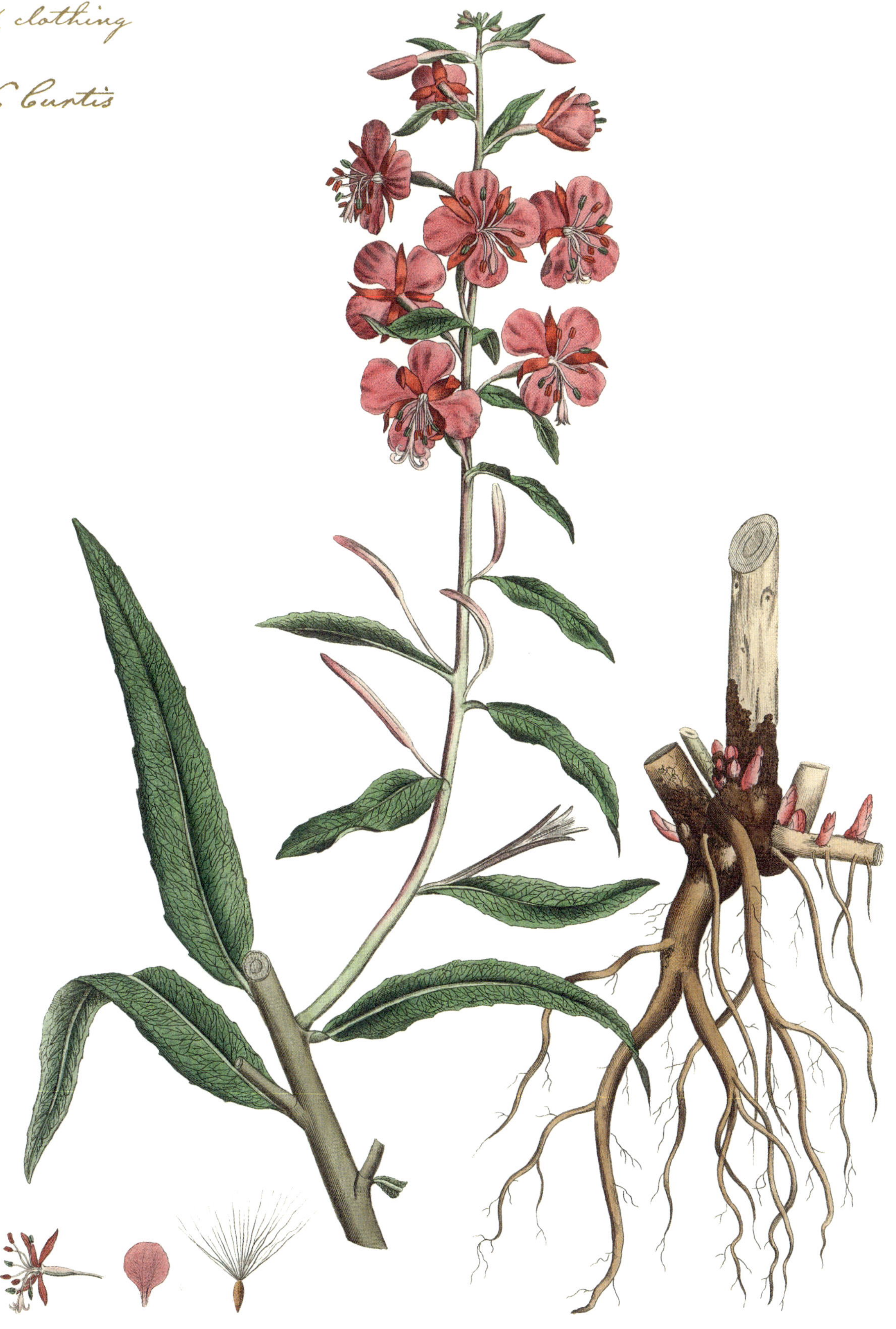

Epilobium angustifolium

MILKWEED

Euphorbia peplus

They say there's no use crying over spilt milk. You might shed a few tears, however, over spilt milkweed. When its stems are snapped, it releases a milky sap that's toxic enough to burn and blister your skin. Even so, herbalists have long made the most of its caustic properties to treat a host of skin disorders, including warts. And scientists today are finding its sap effective against some types of skin cancer.

ANATOMY NOTES

FLOWERS — Easily mistaken for flowers, the eye-popping lime-green flowery growth is actually a collection of bracts, modified leaves that surround, and are often more visually arresting than, the true flower. Here, the real flower is a tiny green speck that itself looks like a little leaf, newly forming.

SEEDS — A fleshy appendage on the seeds works as a kind of handle for ants to hold, helping them disperse the plant. Once in the ground, the seeds can survive for up to 100 years before germinating.

STEM — The toxic sap was used in Kenya to poison arrowheads for hunting.

LEAVES — You can boil the leaves and make a spray for killing slugs and snails.

BOTANIST'S REMEDY

WART TREATMENT — Pumice the top of the wart to break the outer layer, and dab on the sap of a broken milkweed stem, trying not to get it on the surrounding skin. It may sting a little. Remember not to touch your eyes if treating a finger wart. Cover with a plaster or dressing for a week, repeating if necessary. Stop if there's any irritation and seek the advice of a health professional.

ESSENTIALS

Aka — Petty spurge

Family — Euphorbiaceae (spurge)

Urban habitat — Shaded pavement cracks and garden beds

Height — 20 cm / 8 in

Flowers — April to October

Uses — Medicinal, poison

Grow — Sow in spring

Harvest — As required but always wear gloves

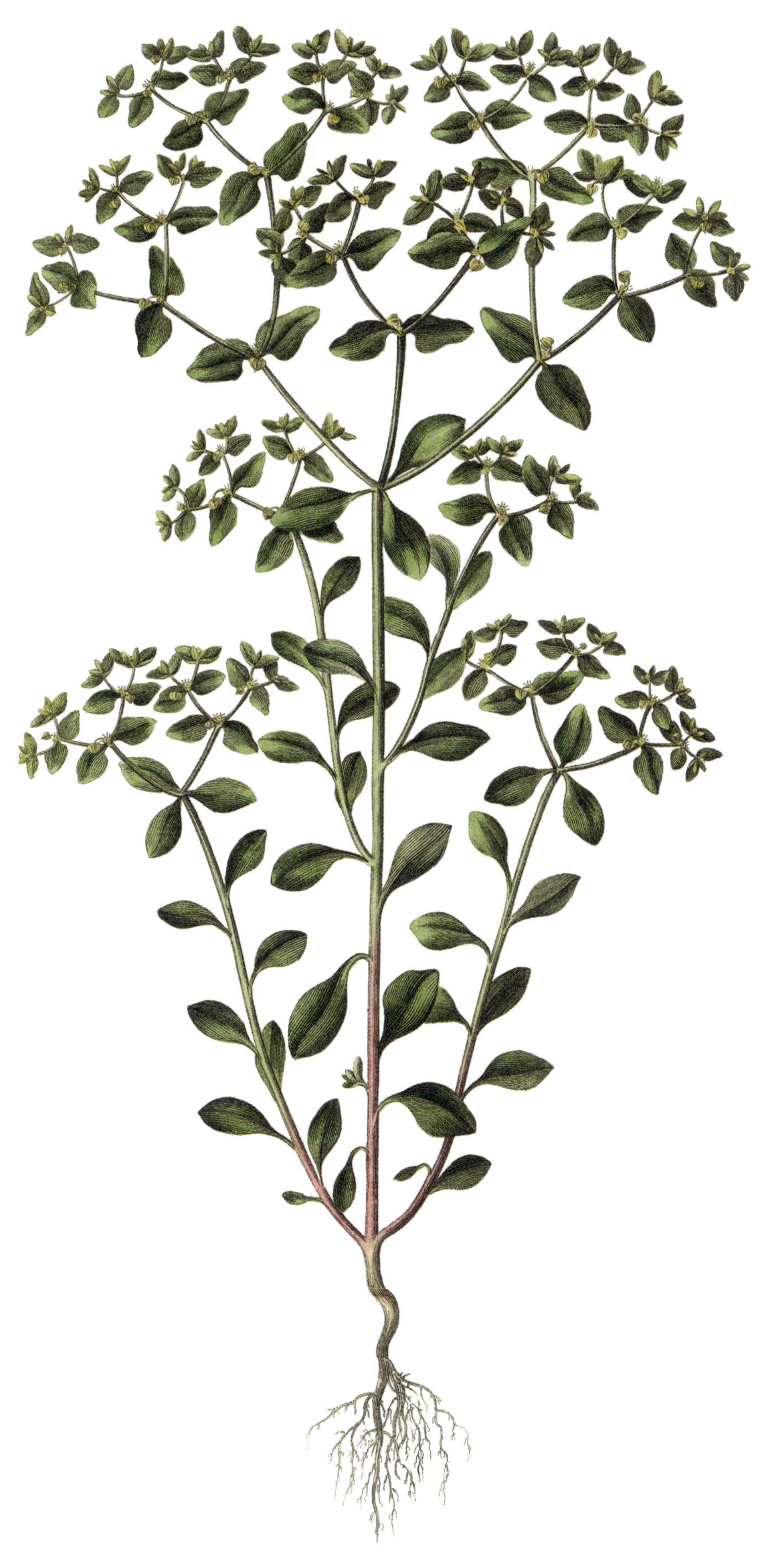

Euphorbia peplus

COMFREY

Symphytum officinale

Comfrey's taproots dig deep into the subsoil, reaching nutrients inaccessible to its competitors. With that extra fuel, it grows quickly, crowding out waterside pathways with bulky masses of hairy leaves, starred with drooping pink-blue flowers. That boost also makes it a highly potent plant, with cell-stimulating properties used for centuries to treat all manner of injuries, even broken bones (hence its folk name, knitbone). It's prized by organic gardeners as a plant food rich enough to satisfy hungry crops like tomatoes and aubergines. Quite the wonder-plant!

ANATOMY NOTES

SEEDS — Comfrey produces large quantities of seeds and spreads quickly. To avoid this in your garden, try planting the naturally hybridised 'Bocking 14', which has sterile flowers.

LEAVES — The leaves make a great, if slightly pungent, plant food. Harvest them as required (they rapidly grow back), cut up and place in a lidded container. Put a weight on the leaves to press the goodness out of them, and cover. In three weeks, there will be a concentrated liquid to add to your watering can, 1 part of comfrey to every 10 of water.

ROOTS — If anything, the roots are even more potent than the leaves. For centuries, they were grated up and applied to fractured bones, helping them to heal and set quickly.

BOTANIST'S REMEDY

COMPRESS — To make a compress for bruises, strains, sprains and joint pain, boil the leaves and soak a flannel in the liquid. Apply the compress, either warm or cold to the affected area.

ESSENTIALS

Aka — Bruisewort, knitbone

Family — Boraginaceae (borage)

Urban habitat — Riverbanks, ditches, waterways, verges

Height — 1 m / 3 ft 3 in

Flowers — May to July

Uses — Heals wounds and fractures, plant food, animal fodder

Grow — Plant out in autumn

Harvest — Pick young leaves in spring

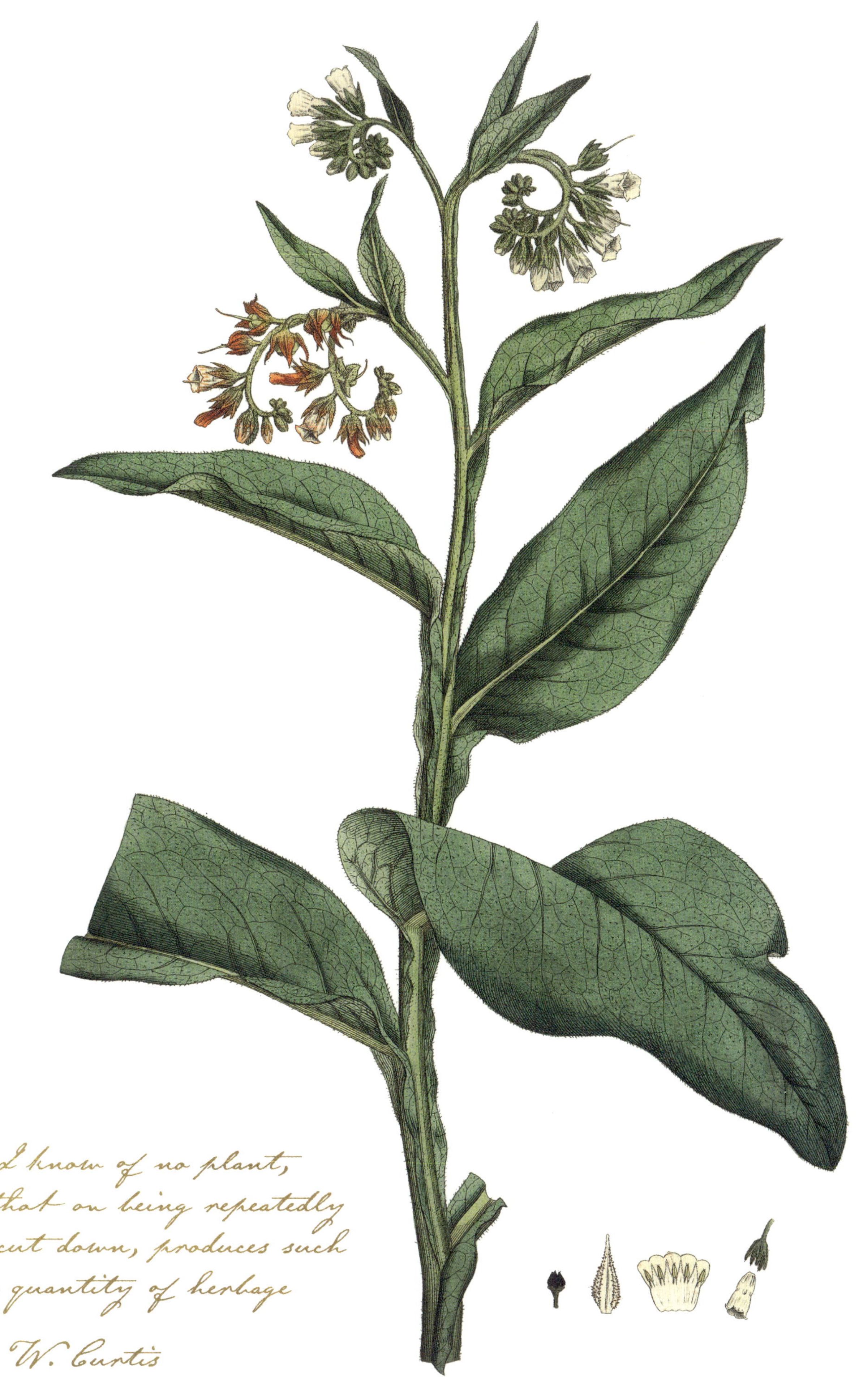

Symphytum officinale

ST JOHN'S WORT

Hypericum perforatum

Ancient herbalists often intuited the use of a plant by its looks. With its profusion of happy yellow flowers, summer sunshine on a stalk, they were right to imagine St John's wort might relieve a case of the blues. Modern research has revealed that St John's wort tea can be as good as antidepressants for treating mild to moderate depression. The plant is also an antibacterial and anti-inflammatory, great for treating knocks, sprains, bruises and sunburn. St John's wort comes into exuberant flower around the summer solstice and its namesake St John's Day (24th June), when it was traditionally gathered to decorate the home and keep evil at bay.

ANATOMY NOTES

FLOWERS — Look closely at the petals and you'll see tiny black dots that look like holes, or the perforations of its scientific name. They're actually small black glands that exude a distinctive scent and red juice when crushed, effective for repelling insects and grazing animals from eating the plant. Mead, a traditional alcoholic honey drink, was sometimes flavoured with the flowers for extra bitter-lemony depth.

LEAVES — The leaves also contain glands, which are best seen when held up to the light, when hundreds of translucent dots suddenly appear. Extracts of St John's wort contain both leaves and flowers.

STEM — The stems are red-tinged and flat-edged. At every node, the point from where a leaf grows, the direction of the stem twists by 90 degrees.

BOTANIST'S REMEDY

TEA — Drinking a tea of 2 or 3 teaspoons of freshly picked St John's wort flowers can help lift your spirits on occasions when you're feeling low or anxious, especially if taken regularly. You can sweeten it with chamomile flowers or honey. Be aware that St John's wort can interact with many commonly prescribed medications (including oral contraceptives), so do get medical advice before using it.

ESSENTIALS

Aka — Devil chaser

Family — Hypericaceae (St John's wort)

Urban habitat — Dry, sunny spots, grassy patches, disturbed ground

Height — 80 cm / 2 ft 8 in

Flowers — May to August

Uses — Treatment of depression, wounds, sunburn; mead ingredient

Grow — Sow seed in spring

Harvest — Flowers in summer

The common people in France and Germany gather it as a certain charm and defence against storms, thunder, and evil spirits

W. Curtis

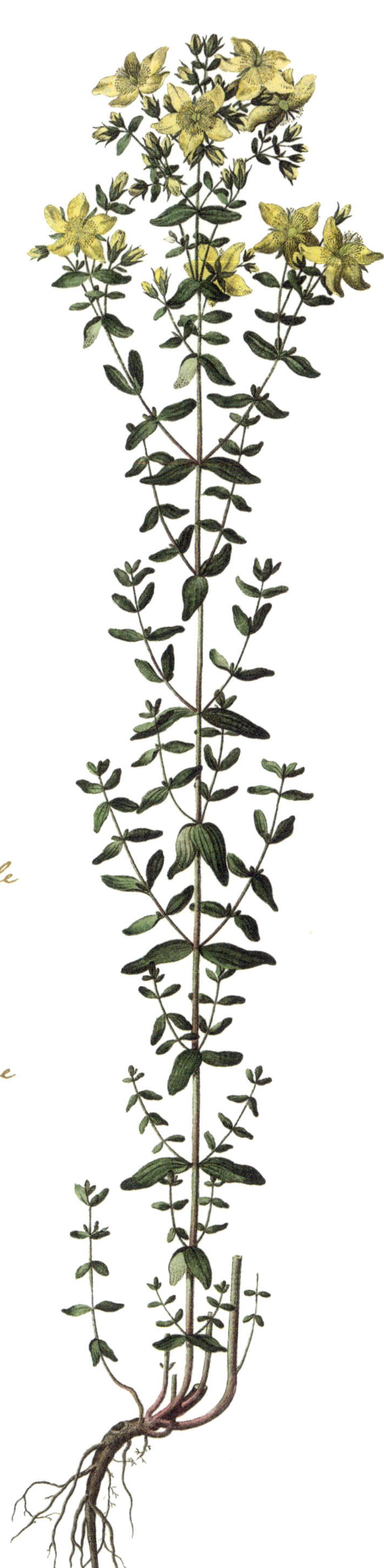

Hypericum perforatum

HERB ROBERT

Geranium robertianum

Liking quiet, shady corners, herb Robert stays low, swiftly covering ground – and occasionally cracked pavements – with dark feathery leaves, dotted with bright pink flowers in summer. As autumn draws in, the whole plant becomes tinged blood-red maroon, fitting for a healing herb long associated with blood and redness, used to stem bleeding, staunch nosebleeds and treat bruising. It may get its name from the 8th-century saint, Rupert of Salzburg, or Robin Goodfellow, better known as the mischievous russet-coloured sprite Puck (made famous by Shakespeare) of midsummer, the time of year when the herb comes into bloom. But it's probably just from *ruber*, the Latin for red.

ANATOMY NOTES

FLOWERS — The flowers are a food source for butterflies and moths, including the scarce barred carpet moth.

LEAVES — The leaves are covered in hair glands that release this herb's distinctive and not very pleasing smell, described as anything from burnt rubber to rotten eggs. In its favour, the nose-wrinkling scent repels mosquitoes, while crushed leaves rubbed on insect bites soothe itching. A poultice helps to reduce bleeding, bruising and encourage healing. A brown dye can be produced from the leaves and other parts of the plant.

SEEDS — Its explosive seedpods catapult seeds up to 5 m / 6 ft away from the parent plant, which can make it something of a runaway rogue in gardens, often considered as a weed.

BOTANIST'S REMEDY

THROAT GARGLE — A mouthwash made from herb Robert can help to soothe a sore throat. Tear up a handful of young leaves and steep them in hot water for 10 minutes before straining. Once the infusion is cool, you can gargle it, repeating three times a day. You could also chew fresh leaves for the same purpose, although the taste is a little overwhelming.

ESSENTIALS

Aka — Death-come-quickly, stinky bob

Family — Geraniaceae (geranium)

Urban habitat — Shady cracks, under hedges, wasteland and margins

Height — 30 cm / 1 ft

Flowers — June to October

Uses — Astringent, antiseptic, antimicrobial, insect repellent

Grow — Sow seeds in autumn

Harvest — Leaves in spring

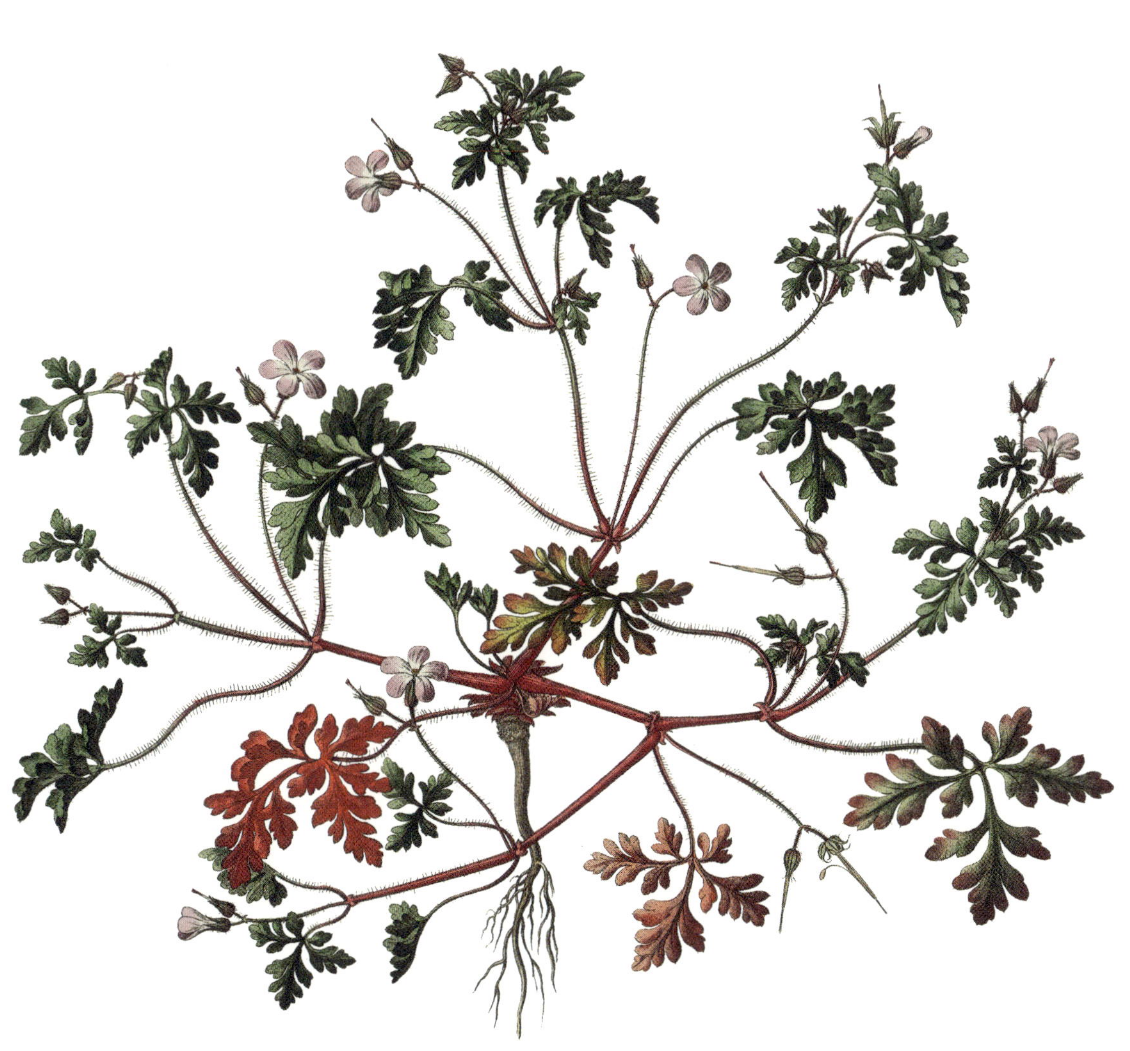

Geranium robertianum

P L A N T A I N

Plantago major

Nothing to do with bananas, this plantain is a useful broadleaved plant often found breaking through trampled soil by footpaths. In fact, it springs up just about wherever humans go, unfurling generous ribbed leaves around spikes of green flowers. Too often overlooked as a wayside weed, it actually has formidable healing qualities, particularly as an antimicrobial that helps stop bleeding and encourages tissue regeneration. The young leaves, lightly blanched, are nutritious in a salad, and even the older stringy ones can fill out a stew. Here's a novelty: a first-aid kit you can eat.

A N A T O M Y   N O T E S

S E E D S — A single plant can produce 20,000 seeds a year, helping it colonise ground quickly. They turn brown when ripe and are protein-rich – handy as a bread topping or ground to add to flour.

F L O W E R S — The flower spikes favour function over beauty. Being wind-pollinated, they don't need to be showy to attract insects.

L E A V E S — In addition to poultices to heal wounds and draw out splinters, the leaves can be made into a balm which relieves the itching from stings. Its herbal tea eases coughs and is particularly flavoursome with a spoonful of honey.

R O O T S — The plant from root to tip brims with therapeutic vitamins and compounds, including antioxidants, antivirals, antidiabetics and antidiarrheals, and has been shown to energise the body and even combat cancer. Who's calling it a weed now?

B O T A N I S T ' S   R E M E D Y

P L A S T E R — Pick a large leaf and wash it. Crush it or chew it up to get the juice flowing and apply to the wound, holding in place with a bandage or clean fabric depending on what is available. Check the wound twice a day, and reapply until it heals.

E S S E N T I A L S

Aka — Cart-track plant, common plantain, greater plantain

Family — Plantaginaceae (plantain)

Urban habitat — Well-trodden paths, roadsides, disturbed ground

Height — 30 cm / 1 ft

Flowers — May to September

Uses — Food, heals wounds

Grow — Sow seed in spring

Harvest — Leaves to eat in early spring, leaves to heal when required

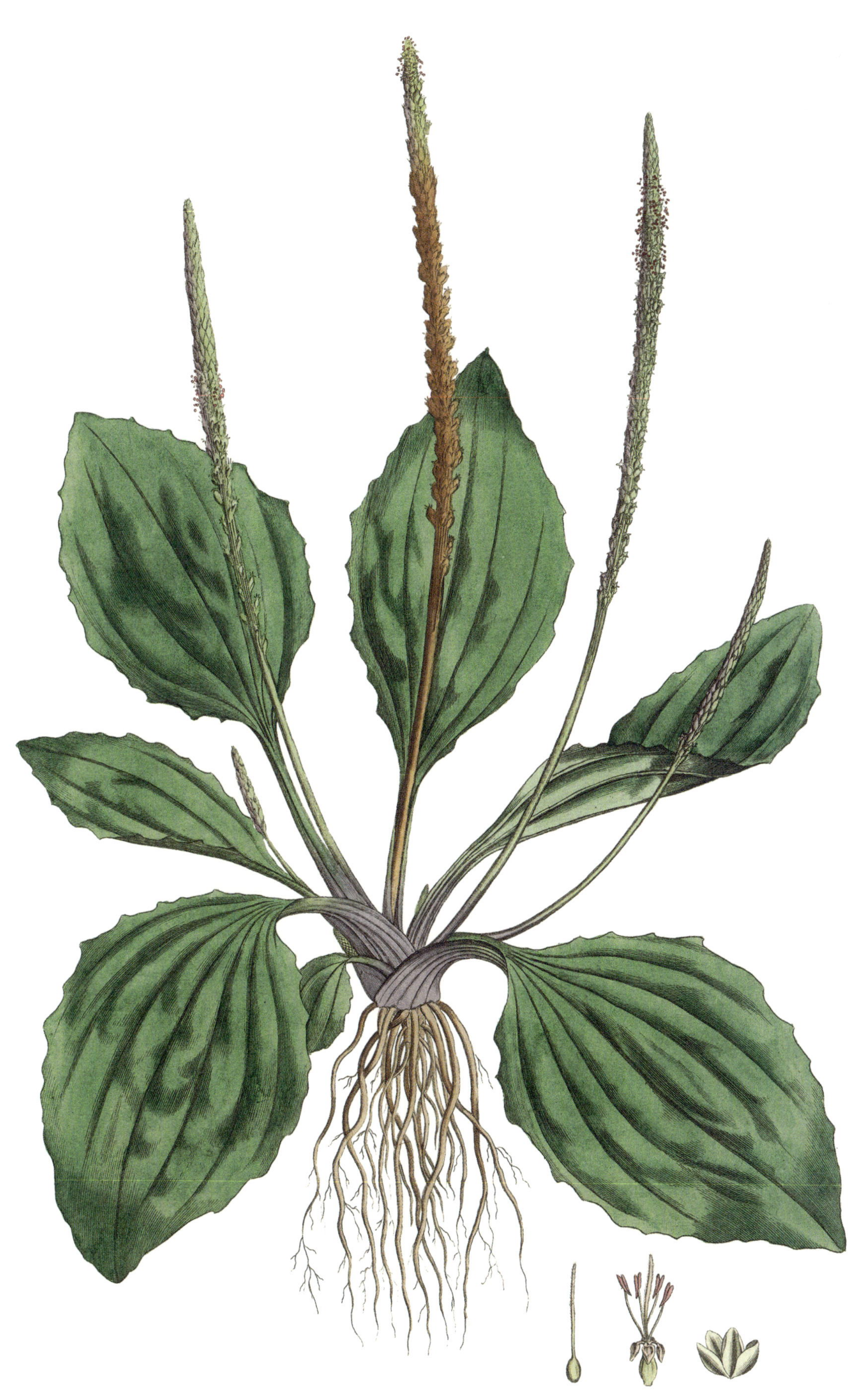

Plantago major

SNEEZEWORT

Achillea ptarmica

Like a practical joker, sneezewort seems to offer its generous white button blooms for a deep scent-filled inhalation. What you'll get in return is a nostril-curling blast of pungency that could trigger a sneezing fit. In fact, the plant was once dried and powdered and taken as snuff to induce sneezing, for medicinal rather than comedy purposes. Its smell is so sharp that it's effective as an insect repellent. It's also quite handy for healing wounds, and gets its name from the great Greek warrior, Achilles, who was told of the plant's healing properties by his tutor, a centaur – while *ptarmica* is the Greek for 'sneezy'.

ANATOMY NOTES

FLOWERS — Some bridesmaids believe that sneezewort in their bouquet will ensure a happy life – if not a sneezy wedding. Flowerheads can also be infused in chilled water for a refreshing summer drink. Hoverflies swarm to them for their nectar.

LEAVES — In the 17th and 18th centuries, leaves were dried and powdered to make a sneezing powder snuff, used to treat migraines and stomach complaints. You can add young leaves sparingly to a salad to give it a bitter edge.

ROOTS — Chewing the fresh root relieves toothache, while root tea is a remedy for tiredness and headaches. The herbalist Nicholas Culpeper said that the fresh root 'evacuates the rheum', helping to clear up discharge from the nose and eyes.

BOTANIST'S REMEDY

INSECT REPELLENT — For this natural repellent, tear up bunches of sneezewort flowers and leaves (homegrown is best), and place into a jar filled with vodka for 3 weeks, shaking now and again. Strain and pour into a small spray bottle. You can add lavender oil for fragrance. Apply as necessary to exposed skin, but do test it on a small area, in case of sensitivity.

ESSENTIALS

Aka — Bastard pellitory, bachelor's buttons

Family — Asteraceae (daisy)

Urban habitat — Wet and boggy ground, and watersides

Height — 60 cm / 2 ft

Flowers — July to September

Uses — Insect repellent, sneezing powder, toothache salve

Grow — Sow seed in autumn

Harvest — Flowers in summer

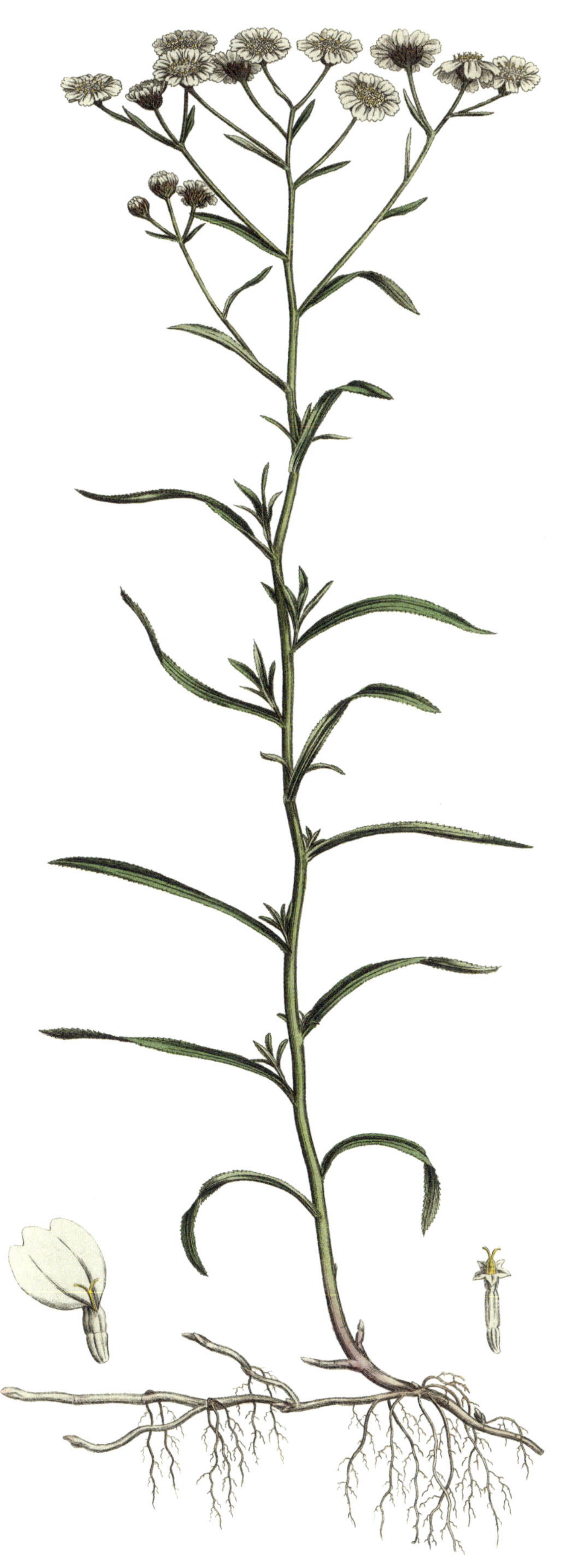

Achillea ptarmica

SELFHEAL

Prunella vulgaris

In neglected grassy patches, selfheal can create sumptuous summer carpets of purple flowerheads for bees to feast on. But that's not the best thing about it. As its name suggests, the true value of this plant is in its extraordinary healing powers. As an astringent, antibacterial and emollient, it well deserves its reputation as one of the best wound herbs (or vulneraries) there is. On top of that, it's great for sore throats, protects the liver, and its antiviral properties are even used to treat cancer. A beautiful and hugely practical little herb to have at hand.

ANATOMY NOTES

FLOWERS — The flowers, which are usually purple but occasionally throw up a white bloom, are rich in nectar, supporting bees, butterflies and wasps. They vaguely resemble open mouths, a possible reason that early herbalists tried using them to cure oral ailments.

LEAVES — A poultice made from crushed leaves and held in place with a bandage can help staunch bleeding, keep wounds clean and encourage healing. You can also eat tender leaves in a salad or add them to chilled water for flavour, which is refreshingly herbaceous.

STEM — An olive-green dye can be extracted from the stems and flowers.

BOTANIST'S REMEDY

THROAT GARGLE — Use this twice a day for sore throats, mouth ulcers and as a general cleansing mouthwash. You can use fresh or dried herbs; plants gathered in summer can be dried for use year-round. Steep 2 tablespoons of flowers and leaves to 250 ml / 8 fl oz of boiled water for 10 minutes. Cool, strain and store in the fridge until use. It should keep for 3 days, or freeze it into ice cubes to suck as needed.

ESSENTIALS

Aka — Woundwort, heal-all

Family — Lamiaceae (mint)

Urban habitat — Grassy parks, verges, paths and lawns

Height — 30 cm / 1 ft

Flowers — July to September

Uses — Healing wounds, medicinal, edible leaves

Grow — Sow seed in spring

Harvest — Leaves for wounds as required; the whole plant in summer for drying

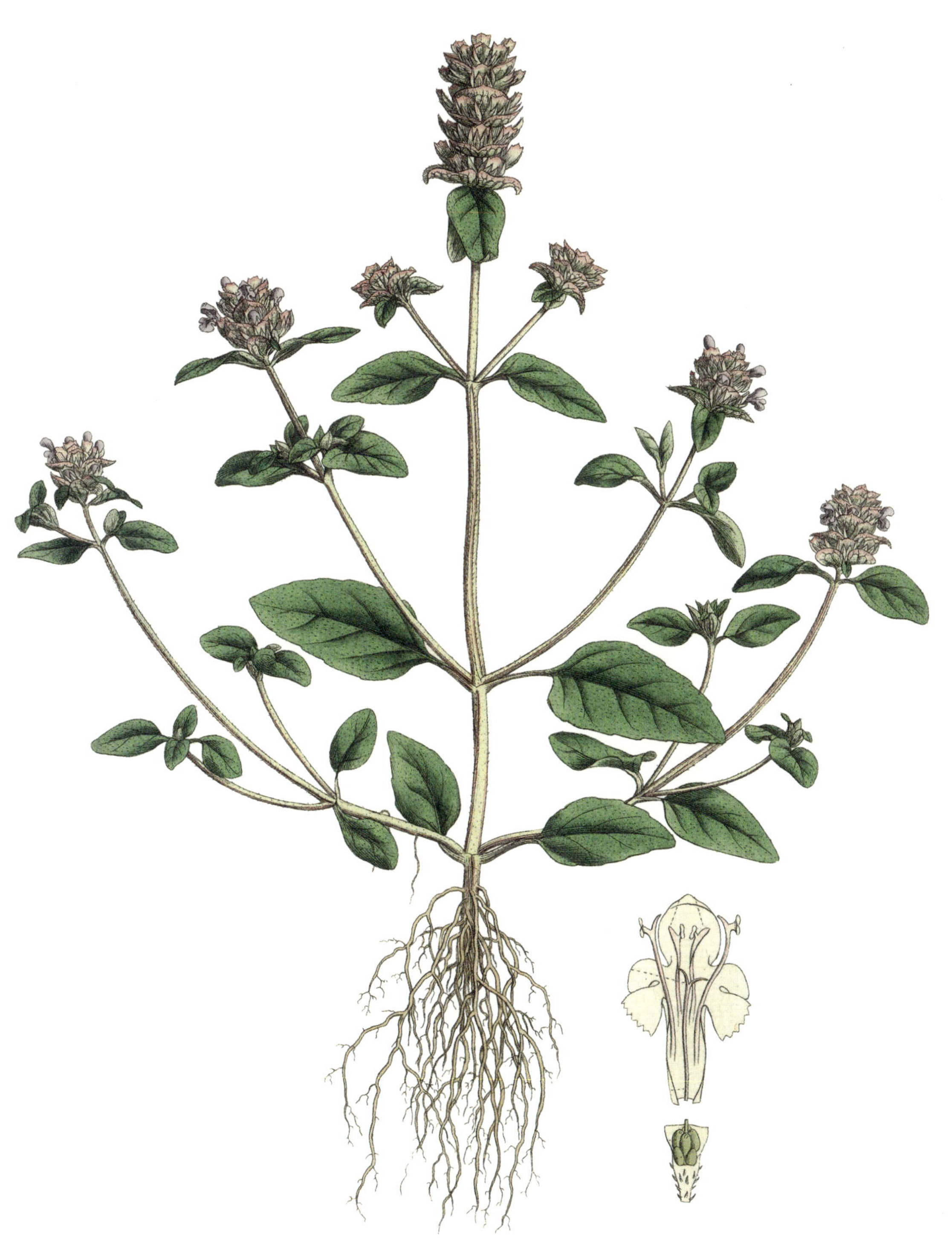

Prunella vulgaris

CONTRIBUTORS

HÉLÈNA DOVE manages the Kitchen Garden at the Royal Botanic Gardens, Kew. Cultivating everything from the allotment staples to more unusual edibles, she is a strong believer in using what grows around us to benefit our daily lives. She is the author of *The Kew Gardener's Guide to Growing Vegetables.*

HARRY ADÈS is the author of *An Opinionated Guide to London Green Spaces* and *A Field Guide to East London Wildlife* with Hoxton Mini Press, and many other titles besides. His two children are far more talented gardeners than he is, and his cats eat the flowers off his only rose bush.

THE ROYAL BOTANIC GARDENS, KEW is a world-famous research organisation and a major international visitor attraction. It harnesses the power of its science, the rich diversity of its gardens and collections to unearth why plants and fungi matter to everyone. It is fighting for a world where nature and biodiversity are understood, valued and protected.

HOXTON MINI PRESS is a small independent publisher based in Hackney, east London. Its small office overlooks a canal – a reminder of how much nature courses through the city. Hoxton Mini Press makes artistic books that are both beautiful and accessible and aim to remind readers of the richness of urban life.

INDEX
OF PLANTS

Achillea ptarmica, 166-67
Adder's fern, 144
Adonis annua, 120-21
Aethusa cynapium, 132-33
African rose, 52
All-heal, 152
Alliaria petiolata, 48-49
Anemone, wood, 130-31
Anemonoides nemorosa, 130-31
Anthriscus sylvestris, 80-81
Apple, thorn, 122-23
Archangel, white, 20
Arctic rattlebox, 84
Arugula, wild, 30
Arum maculatum, 128-29
Asparagus, poor man's, 26
Asplenium scolopendrium, 82-3
Atropa bella-donna, 118-19
Bachelor's buttons (field scabious), 138
Bachelor's buttons (sneezewort), 166
Bairnwort, 146
Balm, bastard, 88-89
Bastard balm, 88-89
Bastard hellebore, 124
Bastard pellitory, 166
Bastard, cress, 22
Bee orchid, 90-91
Bellis perennis, 146-47
Bindweed, hedge, 68-69
Bitter vetch, 46-47
Black nightshade, 126-27
Blessed thistle, 150
Blitum bonus-henricus, 26-27
Blowball, 16
Bombweed, 154
Bouncing Bet, 70
Briar, witches', 74
Broad-leaved dock, 140-41
Bruisewort, 158
Bulbous buttercup, 116-17, front cover
Bunium bulbocastanum, 28-29
Burnt weed, 82
Butomus umbellatus, 104
Butter and eggs, 96
Butter dock, 140
Buttercup, bulbous, 116-17, front cover
Caltha palustris, 40-41
Calystegia sepium, 68-69
Cardamine pratensis, 18-19
Cart-track plant, 164
Chamomile, German, 148-49
Cheese rennet, 36
Cheeses, 56
Chequered daffodil, 78
Chervil, wild, 80
Christ's hair, 82
Clary sage, wild, 42-43
Clematis vitalba, 72-73
Comfrey, 158-59
Common mallow, 56
Common plantain, 164
Common polypody, 144-45
Common teasel, 98
Common toadflax, 96
Conium maculatum, 112-13
Convallaria majalis, 114-15
Corn poppy, 52-53
Cow parsley, 80-81
Cowslip, 136-37
Creeping thyme, 66
Cress, bastard, 22
Crowfoot, wood, 130
Cuckoo flower, 18
Cuckoo pint, 128
Daffodil, chequered, 78
Daggers, 64
Daisy, 146-47
Dandelion, 16-17
Datura stramonium, 122
Dead man's bell, 78
Deadly nightshade, 118
Deadnettle, white, 20-21
Death-come-quickly, 162
Devil chaser, 160
Devil's berries, 118
Devil's vine, 68
Diplotaxis tenuifolia, 30-31
Dipsacus fullonum, 98-99
Dock, broad-leaved, 140-41
Dog poison, 132
Dog rose, 74-75
Earthnut, 28-29
English ivy, 106
Epilobium angustifolium, 154-55
Euphorbia peplus, 156-57
European black nightshade, 126
Featherfoil, 86
Fern, adder's, 144
Fern, hart's tongue, 82-83
Field scabious, 138-39
Filipendula ulmaria, 58-59
Fireweed, 154
Flag, yellow, 64-65
Flowering rush, 104-5
Fool's parsley, 132-33
Fritillaria meleagris, 78-79
Fritillary, snake's head, 78-79
Frog's-stomach, 100
Fuller's herb, 70
Fumaria officinalis, 142-43
Fumitory, 142-43
Galium verum, 36-37
Garden nightshade, 126
Garlic mustard, 48
Geranium robertianum, 162-63
German chamomile, 148-49
Gladiolus, water, 104
Golden maidenhair, 144
Good King Henry, 26-27
Grave myrtle, 62
Greater periwinkle, 62-63
Greater plantain, 64
Green hellebore, 124-25
Guinea-hen flower, 78
Gypsy rose, 138
Hare's lettuce, 50
Hart's tongue fern, 82-83

Heal-all, 168
Heath pea, 46
Hedera helix, 106-7
Hedge bindweed, 68-69
Hellebore, bastard, 124
Hellebore, green, 124-25
Helleborus viridis, 124-25
Hemlock, 112-13
Hen-and-chicks, 94
Herb Robert, 162-63
Hiratake, 44
Honeysuckle, 92-93
Hottonia palustris, 82-83
Houseleek, 94-95
Hylotelephium telephium, 100-1
Hypericum perforatum, 160-61
Iris pseudacorus, 64-65
Iris, yellow, 64
Ivy, 106-7
Jack-in-the-hedge, 48-49
Jimsonweed, 122
Jupiter's eye, 94
Keck, 80
Kingcup, 40
Knautia arvensis, 138-39
Knitbone, 158
Lady's bedstraw, 36-37
Lady's smock, 18-19
Lamium album, 20-21
Lathyrus linifolius, 46-47
Lepidium campestre, 22-23
Lettuce, hare's, 50
Lily of the valley, 114
Linaria vulgaris, 96-97
Lincolnshire spinach, 26
Lonicera periclymenum, 92-93
Lords-and-ladies, 128-29
Lysimachia arvensis, 102-3
Macrolepiota procera, 24-25
Maidenhair, golden, 144
Mallow, 56-57
Malva sylvestris, 56-57
Marigold, marsh, 40-41
Marjoram, wild, 54
Marsh marigold, 40-41
Matricaria chamomilla, 148-49
Mayweed, scented, 148
Meadowsweet, 58-59
Meadwort, 58
Melittis melissophyllum, 88-89
Mercury, 26
Midsummer men, 100
Milk thistle, 150-51
Milkweed, 156-57
Mnium hornum, 108-9
Morel, stinking, 32
Mother-die, 80
Mushroom, parasol, 24-25
Mushroom, pearl oyster, 44
Mushroom, tall, 24
Mushroom, tree oyster, 44-45
Myrtle, grave, 62
Naughty man's cherries, 118
Nettle, 38-39
Nightshade, black, 126-27
Nightshade, deadly, 118
Nightshade, garden, 126
Old man's beard, 72
Ophrys apifera, 90-91
Orchid, bee, 90-91
Oregano, 54-55
Origanum vulgare, 54-55
Orpine, 100-1
Our lady's tears, 114
Oxalis acetosella, 34-35
Papaver rhoeas, 52-53
Parasol mushroom, 24-25
Parsley, cow, 80-81
Parsley, fool's, 132-33
Parsley, poison, 112
Pea, heath, 46
Pearl oyster mushroom, 44
Pellitory, bastard, 166
Pepperwort, 22-23
Periwinkle, greater, 62-63
Petty spurge, 156
Phallus impudicus, 32-33
Pheasant's eye, 120-21
Pignut, 28
Pimpernel, scarlet, 102
Piss-a-bed, 16
Plantago major, 164-65
Plantain, 164-65
Pleurotus ostreatus, 44-45
Poison parsley, 112
Polypodium vulgare, 144-45
Polypody, common, 144-45
Poor man's weatherglass, 102
Poor-man's asparagus, 26
Poppy, corn, 52-3
Primula veris, 136-37
Prunella vulgaris, 168-69
Queen-of-the-meadow, 58
Ranunculus bulbosus, 116-17
Rattle, yellow, 84-85
Red Morocco, 120
Rhinanthus minor, 84-85
Rocket, wild, 30-31
Ropeweed, 68
Rosa canina, 74
Rose, African, 52
Rose, dog, 74-5
Rose, gypsy, 138
Rosebay willowherb, 154-55
Rumex obtusifolius, 140-41
Rush, flowering, 104-5
Sage, wild, 42
Salvia verbenaca, 42-43
Saponaria officinalis, 70-71
Scabious, field, 138-39
Scarlet pimpernel, 102
Scented mayweed, 148
Selfheal, 168-69
Sempervivum tectorum, 94
Shepherd's clock, 102
Silybum marianum, 150
Snake's hat, 24
Snake's head fritillary, 78-79
Snapdragon, wild, 96
Sneezewort, 166-67
Soapwort, 70-71
Solanum nigrum, 126-27
Sonchus oleraceus, 50-51
Sow thistle, 50-51
Spinach, Lincolnshire, 26
Spurge, petty, 156
St Anthony's turnip, 116
St John's wort, 160-61
St Peter's wort, 136
Stinging nettle, 38
Stinkhorn, 32
Stinky bob, 162
Swan's-neck thyme-moss, 108-9
Symphytum officinale, 158
Taraxacum officinale, 16-17
Teasel, 98-99
Thistle, blessed, 150
Thistle, milk, 150-51
Thistle, sow, 50-51
Thorn apple, 122-23
Thunder plant, 94
Thyme-moss, swan's-neck, 108-9
Thyme, creeping, 66
Thyme, wild, 66-67
Thymus serpyllum, 66-67
Toadflax, 96-97
Traveller's joy, 72-73
Tree oyster mushroom, 44-45
Turnip, St Anthony's, 116
Urtica dioica, 38-39
Valerian, wild, 152-53
Valeriana officinalis, 152-53
Vetch, bitter, 46-47
Vinca major, 62-63
Vine, devil's, 68
Violet, water, 86-87
Water gladiolus, 104
Water violet, 86-87
Wax dolls, 142
White archangel, 20
White deadnettle, 20-21
Wild arugula, 30
Wild chervil, 80
Wild clary sage, 42-43
Wild marjoram, 54
Wild rocket, 30-31
Wild sage, 42
Wild snapdragon, 96
Wild thyme, 66-67
Wild valerian, 152-53
Willowherb, rosebay, 154-55
Witches' briar, 74
Wood anemone, 130-31
Wood crowfoot, 130
Woodbine, 92
Woundwort, 168
Yellow bedstraw, 36
Yellow flag, 64-65
Yellow iris, 64
Yellow rattle, 84-85

BENEFACTORS
18TH CENTURY

(A selected few)

Her Grace the Duchess Dowager of Athol, near Farnham, Surrey
Mr Stanesby Alchorne, Tower
Mr Thomas Armiger, Surgeon, Old Fish Street
Mr John Aikin, Surgeon, Warrington
Captain Anningson
The Right Honourable the Earl of Bute, South Audley Street
Sir Joseph Banks, Bart, Soho Square
Sir Lambert Blackwell, Bart. Enfield
Mr Uriah Bristow, Apothecary, Clerkenwell Square
Rev Richard Bluck, Cambridge
Edmund Bott, Esq, Hampshire
John Baker, Esq, Princes Street, Spitalfields
Rev Mr Bagshaw, Bromley, Kent
Rev Nicholas Bacon, Coddenham, Suffolk
Mr William Bent, Clerkenwell
Mrs Brown, Norwich
Mr George Hollington Barker, Attorney, Birmingham
The Right Honourable the Earl of Clanbrassil
Mr Charles Combe, Apothecary, Bloomsbury Square
Mr Loftus Clifford, Surgeon, Mansfield, Nottinghamshire
The Honourable Baron T. Dimsdale
Mr Downing, Surgeon, Clapton
Mr Philip Deck, Bookseller
Dr Dalling, Derby
Mrs Egerton, Oulton Park, Cheshire
Mr Field, Apothecary, Newgate Street
Mr William Fothergill, Yorkshire
Mr Francis Freshfield, Colchester
Mr William Fowle, Apothecary, Red Lion Square
Major Ferrand
Honourable Mr Greville
Captain Gossip
Right Honourable Lord Howe, Grafton Street
Lady Harris, Finchley
Mr Thomas Home, Peckham
Mr W. Henry Higden, Manchester Buildings, Westminster
Rev Robert Harpur, British Museum
Dr Hairby, Spilsby, Lincolnshire
Leonard Troughear Holmes, Esq, Isle of Wight
Right Honourable Lady King, Dover Street, Piccadilly
Right Honourable Lord Loughborough, Lincolns Inn Fields
Rev John Lightfoot, Uxbridge
Abraham Ludlow, M. D. Bristol
Mr Charles Lightfoot, Surgeon, Whitby
Right Honourable the Earl of Marchmont, Mayfair
Right Honourable James Stewart Mackenzie
Sir William Musgrave, Bart, Arlington Street, Piccadilly
Edward Mussenden, Esq
Rev Mr Mills, Norbury, Derbyshire
Captain Manly, Woolwich
Major Morgan, Litchfield
His Grace the Duke of Northumberland
Right Honourable the Earl of Northington
Dr William Newcome, Bishop of Waterford
Mr Robert R. Newel, Surgeon, Colchester
Mr Nisbett, Surgeon, Great Marlborough Street
Her Grace the Duchess Dowager of Portland, Privy Gardens
Right Honourable the Earl of Plymouth, Bruton Street
Honourable Mrs Pitt, Arlington Street
Sir James Pennyman, Bart. Park Street, Westminster
Mrs Petit, Great Marlborough Street
Mr Giles Powell, Apothecary, South Audley-ftreet
Major Thomas Pearson
Peachy, Esq, Wimpole Street
Thomas Ruggles, Esq, Cobham, Surry
Cornelius Rodes, Esq, Barlborough Hall

BENEFACTORS
21st CENTURY

Timothy and Dawn Adès
Fiona Ainsworth
Susan Malberg Albertsen
Sue Albrow
Wendy Alders
Megan Amis
Glenn (Boris) Anderson
Susan Andrews
Ursula Armstrong
Josefin Arneskog
A. Jane Asser
Frank Avocado
Stefanie Bacchi-Andreoli
Patricia Backley
Professor Matthew Bailey
Sophie McKenzie Baker
Jennifer Barnaby
Carroll Barry-Walsh
Connie Barton
Chiara Ugo Baudino
Andrew, Laura and Raphael Beaumont
Simon Beckerman
Laurence Benkhabeb-House
Keith Beven
Jo Bewick
Cornelie de Blécourt
Eleonore de Bonneval
Fiona and Gordon Bow
Jessica Botwright
Kirsten Bradbury
Oscar Brewer
Rob Bridgett
Emilie Broquet
Victoria Brown
Henry Perkins Brown
Jane Brown
Gillian Buckingham
Nick and Robin Bumstead
Valerie and Eric Bumstead
Liangyin Bunsness
Clare Burgess
W. Steven Burke
Sarah Burns
Sue Buss
Kimberly Cameron
Jan and Nathan Cameron
Matthew Caville
Ya-Heng (Judy) Chen
Yulien Chen
Zara Chivers
Paul David Clare
Tim Clements
Jude Clements
Eleanor Coate
Christopher Cockel
Neil P. Confrey
Rory Cooper
Oliver Copeland
Andrew James Cottrell
Haley Cassidy Crocker
Pauline Crogan
Jonathan Crown
Erica Cruse-Jungling
Karen Curzon
Angie Curbill
Rachel Dalton
Heather Deacon
Corinna Isabella and Luigi Del Debbio
Michael Diebold
Eileen Dobson
Helen Dodd
Don and Nick
Derek Dove
Gillian Dove
Annabel and Crispin Downs
Angela Drinkall
Marc Dubois
Roger Duckworth
Susan Dulcie
Liss and Theo Duncan
Knut Engelbrecht
Chiara Faliva
Clarky and Everly Fenn
Heather P. Figueiredo
Sue Grayson Ford
Tom Ford and Jessica Lazar
John Francis
Natasha Frootko
Deborah Frost
Kiyomi Fujii
Gina Fullerlove
Andrew Fullerton
Julia Fullerton
Verity Fullerton-Smith
Jason Gallimore
Eleanor Gavienas
Matt Gleeson
Alice L. Gleeson
Gina Glover
Sally Godstone
Jarnie Godwin
Kate and Pedro Gomes
Lotte Good
Jane Goodall
Nancy Goodwin
Penelope Gresford
Lucy Grey
Vic Grimshaw
Luisa Hailes
Tom Handley
Alastair Hanson
Elisabeth Henn-Carlson
Emerson Grier Hilliard
Liz Hingley
Adam Hinton

Greg Hitchcock
Helen Dorothy Hodge
Giles Hopgood
Rita Hostler
Elizabeth Ingham
Doug Irvine
Flora Ito
Matt Jackson
Sue Jackson
Nicholas Jackson
James
Lisa Johnson
Jane Johnson
Alison Jones
Joshua A. Jones
Dev Joshi
Jeremy Julian
Juho Kajava
Sarah Keegan
Geoff Keene
Torsten Kerler
Ayesha Khan
Emily King
Boguslawa Kowalski
Barbara Krengel
Tingyi Lai & Gueorgui Tcherednitchenko
Françoise Lardet-Veve
Caroline and Michael Lazar
Loveday Lewthwaite
Amanda Lind
Kate Linden
Jo Longhurst
Lois Lovedee
Barbara Mackinder
Andrew Macnaughton
Eula Malenfant
Ellen Louise Manchester
Marcus Marquardt
Sharon McAllister
Tarin McAllister
Kristine McCarthy
Don McConnell
Graham McClelland
Jill McGregor
Sandy McKinnon
James McLaren
Caroline McNulty
Margaret Mee
Jane Mehta
Maren Chumley Michel
Molly Miller-Petrie
Alvaro Carmena Montalvo
Alison Morgan
Kate Rose Morley
Gemma Moss
Paulina Mustafa
Michelle Mylonas
May Nel
Marion Obar
Ben Oliver
Setsuko Ono
Emily Oram
Melissa O'Shaughnessy
Lauriann Owens
Robin Panrucker
Sarah Parrish
Joe Parry
Matthew Peck
Karen Pelham
Matteo Perez
Caille Peri
Dr David Phillips
Rob Phillips
Maggie Pinhorn
Elin Pinnell
Joanne Poulton
Jackie Power
David Praill
Pamela Preene
Megan Prudden
Nuala Quirk
Simon Reece and Wayne Wreford
David J. Rees
Stefanie Reichelt
Fanny Rico
David Rix
Martyn Rix
Shell Roach
Jenifer Roberts
Joanne Roberts
Gary Robinson
Lynn Marie Robinson
Simon Robinson
Todd Riddiough Robinson
Diana Margot Rosenthal
Kay Rossiter-Base
Polly Ruffman
Anna K. Sagal
Jo and Ben Samuel
Heather and John Samuel
Neil and Viv Sanders
Nina Marie de Sanders
Nicky Sandford-Richardson
Lisa Saper
Carol Sargent
Laura Sbrizzi
Catriona Scott
Julia Seavill
Camila Simas
Catherine Simmonds
Nigel Simmonds
Vanessa Simpson
Joe Skade
Paul Smith
Haein Song
Joanne Spiller
Lawrence Spivack
Shannon Spurlock
Claire Stanley
Lorraine Stevenson
Val Stevenson
Joanne Stovell
Iain Sturch
Mark Symons
Richard Taylor
Rosemary Taylor
Jonathan J.N. Taylor
Olivia Temple
Gareth Tennant
Herlinde Tiefenbrunner
Gem Toes-Crichton
Steev A. Toth
Johanna Trew
Beatrice, Marie and Evelyn Turrettini
Sam Vale
Jantiene T. Klein Roseboom van der Veer (Trustee of RBG Kew)
Susanna Venables
Diana Wackerbarth
Lois and Michael Waldvogel
Michael Waller
Frances E. Watson
Susan Waugh
Diana Westlake
Matthew Westley
Lydia White
Ben Whitehurst
Steve Whorton
Brett William
Sam Winston
Kristen Wontorra
Robert Wood
Eoin Woods
Bonnie Etta Wootton
Wayne Worden
Katie Wright
Sophie Wright
Jodie Yates
Adele Yearsley
Malinda Yu

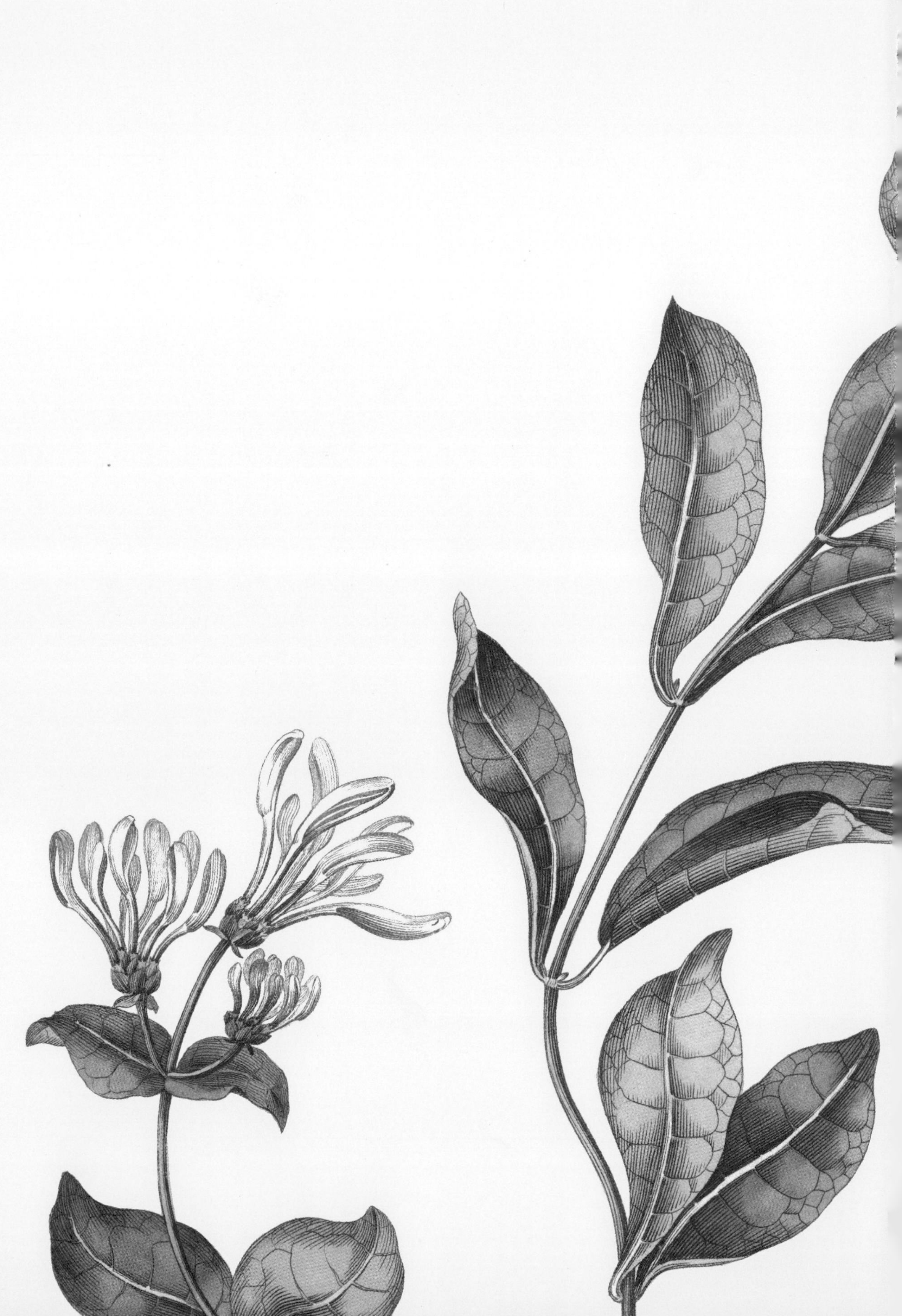